PSYCHE'S RESPONSE TO *Singularity*

CAROLLE M. DALLEY

Dalley Publishing

Paperback ISBN: 978-0-578-22818-1
Hardback ISBN: 978-0-578-22819-8

PRINTED IN THE UNITED STATES OF AMERICA

Table of Contents

Synopsis

Some of the smartest people living in the twenty first century warn us of a rapidly approaching event, Technological Singularity, which will mark the point in time when Artificial Intelligence surpasses human intelligence. Visionaries such as Bill Gates, Stephen Hawking, Elon Musk and Ray Kurzweil all incline to the view that technological advancements will surpass human ability to control technology, with the result that technology either enslaves or exterminates humans. Ray Kurzweil wrote the book "The Singularity Is Near" to make the danger clear. He predicts that Singularity will occur in the period between 2020 and 2045.

I believe the visionaries warn us of Singularity because they have good intentions, but I think their unwritten assumption is wrong. They assume that technology is evolving at an exponential pace, while humans evolve at a linear pace. In my opinion, technology and humans evolve in tandem. I illustrate my perspective by reference to a series of revolutions that are milestones in human evolution: the Agricultural Revolution, the Scientific Revolution, the Industrial Revolution and the Digital Revolution. Early in each revolution, the human psyche experiences a dethronement from a lofty place of presumed superiority in the universe. The dethronement is followed by an acquisition of new knowledge obtained during the revolution. As the revolution progresses, the newly acquired human knowledge is offloaded to technology. That offloading frees humans to pursue new mental capabilities. In our current revolution, the Digital Revolution, the cause for dethronement of the human psyche is the discovery that Artificial Intelligence is not just able to perform rational functions faster and more accurately than humans, but that Machine Learning can be accomplished without human cooperation. Machines can learn from Big Data. Moreover, Machine Learning occurs in an iterative learning process, that enables the machines to produce generations of software that are increasingly smarter than previous generations.

The introduction of Machine Learning results in a dethronement for the human psyche, because we have long believed that intelligence made us superior to all other creatures in nature. Now, there are inanimate machines that can make themselves smarter without needing us to write software for them to execute. That is a dethronement from the pinnacle of intelligence. I believe that what is occurring now is an offloading of rational capabilities from humans to technology, in the form of Artificial Intelligence. I share digital visionary Kevin Kelly's view that Artificial Intelligence is becoming a utility, that is similar to electricity. I expect Artificial Intelligence to embody rational capabilities and make them continuously available to humans. In my opinion, that will free humans of responsibilities that are strictly rational and enable us to turn our attention to developing our non-rational capabilities. Rational capabilities are those based on logical thinking. Non-rational capabilities are those related to topics such as creativity, feeling, motivation, ethics, values and insight. Non-rational capabilities do not necessarily operate in the conscious layer of the human psyche; they can be latent in the sense that they exert their influence from the unconscious layer of the psyche. After offloading rational capabilities to Artificial Intelligence, I believe that humans will have greater opportunity to bring into conscious awareness those mental capabilities that operate in the unconscious layer of the psyche.

My prediction is that the current Digital Revolution will be followed by a Psychological Revolution, in which Analytical Psychology will re-emerge and acquire mainstream acceptance. When Carl Jung introduced Analytical Psychology in the early twentieth century, it did not gain mainstream acceptance because it was not considered a science. Its principles could not be proven or falsified. At the time, there were no tools suitable for making Analytical Psychology a science. I believe we now have the tools. Technology offers tools in Artificial Intelligence, Machine Learning and Deep Learning. Artificial Intelligence can mimic the rational capabilities of the psyche. Machine Learning can conduct iterative learning from psychological data and improve its knowledge continuously. Deep Learning can use natural language processing to detect patterns in ordinary language, whether written or spoken. Analytical Psychologists now have the opportunity to give their discipline a scientific footing. In the upcoming Psychological Revolution, I believe Analytical Psychology will re-emerge to play a significant role because, unlike many other schools of psychology, it pays attention to the psyche as a whole. Attention to the whole psyche is becoming increasingly important. As we offload rational capabilities to technology, our future will depend on how well we get to know our non-rational capabilities and bring them into consciousness.

I agree with historian Yuval Noah Harari that intelligence is being de-coupled from consciousness and being relegated to machines. Artificial Intelligence does not have consciousness, so humans are still ahead in terms of capabilities of the psyche. We now need to expand our consciousness by bringing into it those capabilities that are latent in the unconscious part of our psyche. As Artificial Intelligence becomes a utility, we will have less need to focus our resources on activities requiring rationality. Already, Artificial Intelligence provides systems for rational activities such as monitoring home security, driving cars by intelligence built into the dashboard, and supporting human organs by electronic devices containing built-in health information.

To explain my prediction that the Digital Revolution will be followed by a Psychological Revolution, I rely on the psychological principles of enantiodromia and projection. I use those principles to explain the underpinnings of the upcoming Psychological Revolution. Enantiodromia informs us that when conscious attention is focused on an extreme, it precipitates a counter movement that emerges from the unconscious psyche. Singularity is an extreme in the focus of conscious attention on rationality. It will result in a counter movement away from rationality, toward non-rational functions of the psyche. As Singularity approaches, we are already offloading much of our rational capabilities to algorithms and electronic devices. When we are unencumbered by rational activities, we can pay more attention to expanding our consciousness to include functions that have been latent in the unconscious aspect of our psyche. That will put us in a position to take advantage of the best of Artificial Intelligence, while exploring the undiscovered potential of our human intelligence.

I make the following predictions to summarize Psyche's Response to Singularity:

- Technological Singularity will not result in the enslavement of humanity by 2045, because technology focuses on rationality, which is only a fraction of the capabilities of the human psyche. I use the principle of psychological projection to explain that the psyche and technology have a history of evolving in tandem.
- The current Digital Revolution will be followed by a Psychological Revolution. I use the psychological principle of enantiodromia to explain that there will be a counter movement to technology's extreme focus on rationality.
- Analytical Psychology, which never achieved mainstream acceptance, will re-emerge during the Psychological Revolution, to acquire a scientific footing. I use the principles of Artificial Intelligence, Machine Learning and Deep Learning to explain the technology now available to support the achievement of a scientific footing.

I challenge psychologists to lead an initiative that shapes the future of our psychological growth by taking advantage of the available technology.

Preface

The theme of this book is that Technological Singularity is not inevitable because technology and the human psyche are both subject to the psychological principle of enantiodromia. Enantiodromia is a principle which states that when an extreme, one-sided tendency dominates conscious life, an equally powerful unconscious counter action builds up, first to inhibit the conscious tendency, then to break through the conscious control. The term "enantiodromia" depicts the interplay between conscious and unconscious opposites throughout the course of life. When there is an imbalance, enantiodromia restores balance between the conscious and unconscious aspects of the psyche. A restoration becomes necessary when the conscious focus of the psyche approaches an extreme. The restoration is accomplished by the emergence of an unconscious counterpart to the conscious focus.

Those who promote Singularity claim that Artificial Intelligence will surpass human intelligence by the middle of the twenty first century, and either enslave or exterminate humans. I believe that is an extreme view which preoccupies our conscious attention, and I expect an emergence of an unconscious opposite that will be a response to the conscious extreme. I see Singularity as an extreme view because it predicts the enslavement or extinction of the human species. I doubt that technology will surpass human intelligence because Singularity proponents focus on capabilities that they know how to digitize and measure, while the human psyche possesses capabilities far beyond cognitive capabilities. We are now in the second decade of the twenty first century. If Artificial Intelligence is to surpass human intelligence by the middle of the century, it has a lot of catching up to do just to be on par with the psyche.

These are my predictions about the future:

1. Singularity will not become a reality by the year 2045. Artificial Intelligence will not enslave or exterminate humans. Singularity is an extreme view in conscious attention. Based on the psychological principle of enantiodromia, I predict that the extreme view will trigger the emergence of a counter movement from the unconscious psyche.

2. In the unconscious part of the psyche, there is both personal and collective content. In the conscious part of the psyche, there is personal content, but no collective content. I see that as an imbalance. I predict the emergence of a "consolidating mind" which will be a collection of individual minds that function in concert when triggered by events that impact large portions of the global population.

3. I predict that the current Digital Revolution will be followed by a Psychological Revolution. During the Psychological Revolution, humans will become more knowledgeable about the psyche. The psychological principle of projection will be instrumental in understanding how we relate to people and institutions in the external world. The Jungian Typology will help us manage the interior world of our psyche.

4. It has been over a hundred years since Carl Jung introduced Analytical Psychology, and it has not acquired a scientific footing. I predict that Analytical Psychology will re-emerge and obtain a scientific footing, with the aid of Artificial Intelligence.

These are the goals and scope of this book:

- One goal is to explain my prediction of a paradigm shift from the current Digital Revolution to an upcoming Psychological Revolution. The scope of this goal is to demonstrate that the paradigm shift is compatible with the psychological principle of enantiodromia.

- The second goal is to provide examples of the psychological principle of enantiodromia in the history of the Western world. The scope of this goal is to explain the transitions from Religion to Science, from Enlightenment to Romanticism, from Modernism to Postmodernism, and from Empiricism to Unconscious Dynamics.
- A third goal is to explain that Artificial Intelligence provides the tools that can help Analytical Psychology to acquire a scientific footing. The scope of this goal is to define Artificial Intelligence, Machine Learning and Deep Learning as tools that can support Analytical Psychology in its quest for a basis in science.
- A fourth goal is to demonstrate that historically the human psyche and technology evolve in tandem with each other. Since they evolve in tandem, this makes Singularity unlikely and diminishes the fear that technology will enslave humanity. The scope of this goal is to show that the psyche and technology evolve in tandem during four historical revolutions. They are the Agricultural, Scientific, Industrial and Digital Revolutions.
- The fifth goal is to demonstrate that people develop one function of the psyche in early life, then develop latent functions of their psyche after life-changing events in their lives. The scope of this goal is to use brief biographies of Albert Einstein, Coco Chanel, Julia Child and Jeff Bezos to show development of functions in the Jungian Typology.

The audience for this book:

I write for an audience of adults who have a layperson's interest in psychology. Readers need not have any specialized knowledge in psychology or technology. I explain the terms that I use, both in the chapters and in the Glossary.

Arrangement of this book:

The chapters and sections do not all need to be read in the order presented.

- Chapter 1 describes Technological Singularity and identifies what I consider to be flaws in the prediction that Artificial Intelligence will surpass human intelligence.
- Chapter 2 provides a definition of the psychological principle of enantiodromia.
- Chapter 3 through Chapter 6 describe examples of enantiodromia in Western history. Readers who are familiar with the application of enantiodromia to history may wish to skip certain sections.
- Chapter 7 contains a description of basic components of the psyche that are adequate to support a response to the claims of Singularity.
- Chapter 8 offers psyche's response to Singularity in terms of the psychological projections that occurred in historic revolutions. The sections about historic revolutions can be read in isolation of skipped.
- Chapter 9 contains predictions about the future of humanity based on physical sciences. These are predictions of scientists Teilhard de Chardin and Ray Kurzweil.
- Chapter 10 is about my predictions for the future of humanity based on principles of Analytical Psychology and principles of Artificial Intelligence.
- Chapter 11 is about my challenge to psychologists to take the initiative of shaping the future of psychological growth.
- Chapter 12 provides brief biographies of people who developed one function of their psyche in early life, then developed a latent function due to significant events in their lives. Readers who are familiar with the Jungian Typology may wish to skip certain sections.

- Chapter 13 offers perspectives of other authors on the topic of Singularity.

The Cover Art depicts an ongoing interaction between the psychology of humans and the technology of robots. The composition is a metaphor for a mutual relationship. The psyche enriches technology by downloading new knowledge. Technology supports the psyche by taking over the activities that humans find laborious.

Acknowledgments

I appreciate the contribution of Kevin P. Richard, former lecturer in Archetypal Pattern Analysis at the Assisi Institute, USA. He reviewed drafts of this book and provided consultation about the human psyche.

I am grateful to the UPS Store # 4670 for doing the graphic design work that made my hand-drawn diagrams into digital images.

As a first-time author, I am fortunate to have the staff at Outskirts Press support the publication of this book.

Introduction

The expression "winter of AI" caught my attention. I wondered why would Artificial Intelligence (AI) experience a winter. Artificial Intelligence is about advancing technology, it is being pursued by some of the smartest minds of our generation, and it embodies some of the most noble goals of humanity. So, I did some research and found out that AI has had two winters. These winters have a sequence. Over-promise. Under-deliver. Then, hibernate for a winter of reduced funding. Artificial Intelligence has a history of promising more than it can deliver. When it fails to deliver, funding dries up. Then, there is a winter of AI.

Artificial Intelligence has a new promise. This one is being characterized as Technological Singularity. Technological Singularity refers to the prediction that Artificial Intelligence will surpass human intelligence by the year 2045, when it will either enslave or exterminate humans. That sounds to me like an approaching "winter of AI" because Artificial Intelligence has never addressed human intelligence as a whole. To surpass human intelligence, Artificial Intelligence needs to consider all of the capabilities of the human psyche. There are capabilities in conscious attention and there are latent capabilities in the unconscious part of the psyche. Artificial Intelligence focuses on rationality and cognition, which are just parts of human intelligence.

I see an opportunity for Artificial Intelligence and Analytical Psychology to work out a symbiotic relationship. Analytical Psychology can teach Artificial Intelligence how to avoid "winters of AI" by paying attention to the pitfalls of the hubris entailed in over-promising. Artificial Intelligence can help Analytical Psychology to find the scientific footing that has eluded it since Carl Jung outlined his psychology over a hundred years ago.

This book covers the joint evolution of the psyche and technology over the course of revolutions in history. By following their evolutions through the Agricultural Revolution, the Scientific Revolution, the Industrial Revolution and the Digital Revolution, I demonstrate that the psyche and technology evolve in tandem. This joint evolution undermines the claim of promoters of Technological Singularity that Artificial Intelligence will surpass human intelligence. I acknowledge that there is potential for rogue technology, hacking and mistakes, all of which can harm humanity. However, I dispute the claim that Artificial Intelligence will surpass human intelligence by 2045. So far, there is no articulation of what exactly Artificial Intelligence will surpass. There is neither a definition of human intelligence nor a declaration of which components of human intelligence will be surpassed. I foresee an approaching paradigm shift from a narrow focus on rationality to a broader outlook that encompasses both the conscious functions and the latent unconscious capabilities of the human psyche.

The value of Analytical Psychology is not limited to clinical situations, where patients recover from mental illness. Analytical Psychology is also valuable in understanding the civilization in which we live. Technology is a pervasive influence in our current civilization, both at the individual level and the collective level. It is time for psychologists to step up to the responsibility of shaping the future of human psychological growth, in a world where technology is ubiquitous. Psychologists do not have to do that alone. There is an opportunity for a symbiotic relationship between Artificial Intelligence and Analytical Psychology.

In Chapter 1, I explain the nature of Technological Singularity. I also explain the threat to humanity as defined by Ray Kurzweil, a well-known promoter of Technological Singularity. Then, I outline what I consider to be flaws in Kurzweil's arguments.

Chapter 1

———✺———

The Threat of Technological Singularity

This chapter is about the threat that Technological Singularity poses for humans, according to Ray Kurzweil. After quoting Kurzweil's explanation of Technological Singularity, I explain my opinion that Singularity is an extreme in conscious attention. Based on Carl Jung's principle of enantiodromia, Singularity is therefore a precursor to the emergence of a counter movement from the unconscious psyche. I point out that Singularity has something in common with the Age of Enlightenment: a reliance on science as if it can encompass the entirety of knowledge. I agree with Kurzweil's description of the rapid pace of technological development, but I point out what I consider to be the flaws in his argument that humanity will become subservient to technology.

1.1 Kurzweil's Claims about Singularity

In his book "The Singularity Is Near", Kurzweil describes Singularity as a future point in time when rapid technological advances will surpass human ability to control technology, with the resulting threat that humans may be enslaved by technology. Here are quotations from Kurzweil's general predictions about Singularity:

"The Singularity will represent the culmination of the merger of our own biological thinking and existence with our technology, resulting in a world that is still human but that transcends our biological roots. There will be no distinction, post-Singularity, between human and machine or between physical and virtual reality.

... Although Singularity has many faces, its most important implication is this: our technology will match and then vastly exceed the refinement and suppleness of what we regard as the best of human traits." [1]

These are some more specific excerpts from Kurzweil's thoughts about Singularity.[2]

- "By the late 2020s we will have completed the reverse engineering of the human brain, which will enable us to create nonbiological systems that match and exceed the complexity and subtlety of humans, including emotional intelligence."
- "(T)he gradual but inexorable progression of human themselves from biological to nonbiological ... has already started with the benign introduction of devices such as neural implants to ameliorate disabilities and diseases."
- "(T)hese trends continue until our nonbiological intelligence vastly exceeds that of the biological portion."
- "In the 2040s, when the nonbiological portion will be billions of times more capable, will we still link our consciousness to the biological portion of our intelligence?"
- "The issue of who or what is consciousness and the nature of subjective experiences ... are fundamental to our concepts of ethics, morality, and law."
- "We can measure the correlates of subjective experience (for example, certain patterns of objectively measurable neurological activity with objectively verifiable reports of certain subjective

experiences, such as hearing a sound). But we cannot penetrate the core of subjective experience through objective measurements. (W)e are dealing with the difference between third-person 'objective' experience, which is the basis of science, and first-person 'subjective' experience, which is a synonym for consciousness."

In my opinion, Kurzweil's book is a well thought out treatise by a futurist, who has an intriguing vision and who articulates it well. His view is shared by technology experts like Stephen Hawking, Bill Gates and Elon Musk, all of whom endorsed Kurzweil's book. While I agree with the explanation that the rapid pace of changing technology will bring many sophisticated devices to support our lifestyles and improve our health, I do not share their idea that humanity will become subservient to technology.

Technology does have the potential to harm humans. A hacker can use technology to implement bad intentions that harm a society. A technological mistake can result in inaccuracies that setback normal processing of enormous volumes of data. An artificially intelligent system can go rogue and create havoc in an institution. In addition, there are increasing incidents where some humans are simply overwhelmed by technology. However, I do not believe that humanity in general will become servants to technology.

1.2   Flaws in Kurzweil's Argument

Here are my thoughts on the flaws in Kurzweil's argument:

- Kurzweil appears to make an unwritten assumption that the human psyche is not evolving at a pace comparable to the evolution of technology. He writes as if he assumes that technology is evolving rapidly and racing past the human psyche that is without any prospects in evolution.

- He appears to be unaware of the psychological principle of enantiodromia. He seems to be unfamiliar with the history of humanity which shows that when there is an extreme in conscious attention, there emerges a counter movement. For example, the Age of Enlightenment was very similar to Singularity in the sense that both focus on cognitive capabilities and mechanical devices, while dismissing consciousness and its creative sensibilities. Kurzweil does seem to be aware that Singularity is an extreme. That is why he wrote a book to warn us about its threat. However, he does not seem to be aware that an extreme is a precursor to a counter movement from the unconscious aspect of the psyche.
- "The Singularity Is Near" pays a lot of attention to cognition and to neuroscience. That is giving attention to what can readily be measured. Neuroscience has the benefit of imaging technology. For example, Magnetic Resonance Imaging (MRI) machines can produce scans of the brain that can be aggregated to create analyses about brain activity. Those analyses have various uses, such as determining the degree of health of the brain, early detection of diseases and provision of information about the type of activities going on in the brain. Brain scans do not inform us of the motivation, the values, the emotion or the intentions behind the scans. Neuroscience bolsters Kurzweil's argument that technology is advancing, but Neuroscience does not offer enough substance to compare the technological scans with the intelligence of a human mind.
- While claiming that nonbiological systems will match and exceed the complexity and subtlety of humans, including emotional intelligence, Kurzweil fails to take the human psyche into account.
- He acknowledges that he has not addressed what he calls the "vexing question of consciousness". He is aware that consciousness needs to be addressed, but admits not having

addressed it yet. Regarding unconsciousness, he does not appear to be aware that it also needs to be addressed. Kurzweil's book gives the impression that he is unaware that consciousness needs to be addressed in conjunction with unconsciousness. Together they function as a whole psyche. In an invisible manner, the unconscious part of the psyche influences the rationality that goes into the prediction of Singularity.

- To his credit, Kurzweil wrote a later book to address the mind and thought, both of which were given limited attention in "The Singularity Is Near". The title of the later book is "How to Create a Mind: The Secret of Human Thought Revealed" and it is an interesting book, but I found that it tends to treat mind and brain as if they are interchangeable. I do not believe they are interchangeable. After all, the brain is physical while the mind is non-physical.

Kurzweil is credited with having made accurate predictions about technology in the past. His prediction about Singularity is that the middle of the twenty first century is when we should expect technology to surpass humans in intelligence. I am willing to place my confidence in his timing, but not in the content of his prediction. What I am expecting in the middle of the twenty first century is that we will see the effect of the enantiodromia principle. There will be a shift in our outlook. The extreme characteristics ascribed to technology in our conscious attention will give way to an emergence from our unconsciousness. My prediction is that the Digital Revolution will give way to a Psychological Revolution. Technology will continue to advance in sophistication, but it will not dominate our consciousness. In the chapter titled "Predictions Based on Psychological Principles", I provide details about my prediction that a Psychological Revolution will follow the Digital Revolution.

NOTES

1. For general predictions about Singularity, see page 9 of "The Singularity Is Near" by Ray Kurzweil.
2. For specific details about Singularity, see pages 377 - 379 of "The Singularity Is Near" by Ray Kurzweil.

Chapter 2

------ ∾ ------

Enantiodromia: Definition

The Swiss psychologist, Carl Jung, used the word "enantiodromia" to explain a principle by which societies evolve psychologically. He offered the following as a definition of the principle of enantiodromia:

> Enantiodromia is: "the emergence of the unconscious opposite in the course of time. This characteristic phenomenon practically always occurs when an extreme, one-sided tendency dominates conscious life; in time an equally powerful counter-position is built up which first inhibits the conscious performance and subsequently breaks through the conscious control." [1]

To further explain the principle of enantiodromia, I prepared an image to show how the evolution of the psyche proceeds by an interplay between conscious and unconscious opposites. The opposites that I select are the Age of Enlightenment and the Age of Romanticism.

FIGURE 2.1
AN INSTANCE OF ENANTIODROMIA

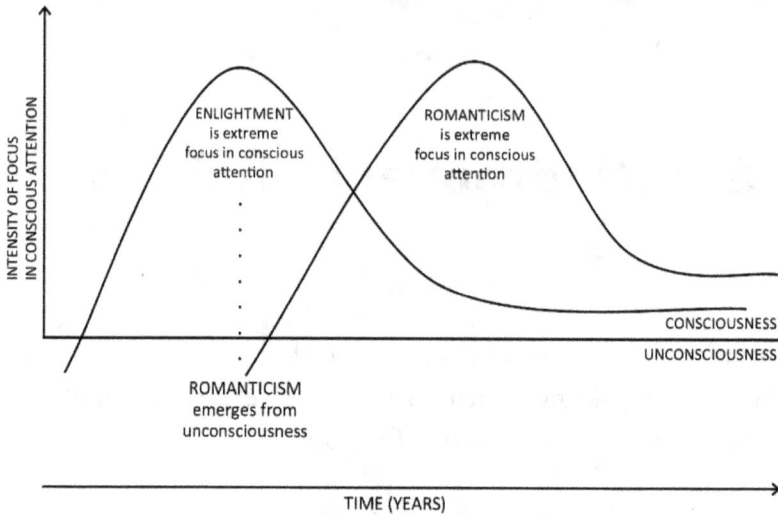

INTENSITY OF FOCUS IN CONSCIOUS ATTENTION

ENLIGHTMENT
is extreme
focus in conscious
attention

ROMANTICISM
is extreme
focus in conscious
attention

CONSCIOUSNESS

UNCONSCIOUSNESS

ROMANTICISM
emerges from
unconsciousness

TIME (YEARS)

Figure 2.1 shows an instance of enantiodromia, where the Age of Enlightenment gave way to the Age of Romanticism over the course of time. The Age of Enlightenment covered a period of time from the seventeenth to the eighteenth century. The Age of Romanticism occurred from the middle of the eighteenth century to the middle of the nineteenth century. In Figure 2.1, the horizontal line with the arrow shows the passage of time in terms of years. The vertical line shows the intensity of focus on topics in the conscious attention of society. The curve on the left is the Enlightenment curve. The curve on the right is the Romanticism curve. The dotted line shows the point in time when the Enlightenment curve has its highest intensity in the conscious attention of Western society. At the point in time when Enlightenment is at an extreme in consciousness, the Romanticism curve begins to emerge from unconsciousness. The Romanticism curve moves from unconsciousness into consciousness. The Romanticism curve shows an increasing intensity and peaks when Romanticism reaches an extreme in conscious attention. Then the curve begins to show decreasing intensity in conscious attention. Figure 2.1 shows an instance of enantiodromia where the Age of Enlightenment gave way to the Age of Romanticism.

An emergence is usually precipitated by a bifurcation that begins to restrain the extreme conscious focus, then over the course of time, develops into a counter movement that outweighs the force of the conscious focus. I provide more information about this bifurcation in the Chapter titled "Enantiodromia: From Enlightenment to Romanticism".

In the next chapter, I provide the first of a series of examples of enantiodromia in the history of Western society: the transition from Roman Catholicism to Newtonian Science.

NOTE

1. For more information about enantiodromia, see the book "Psychological Types" page 426. The Collected Works of C. G. Jung, Volume 6, paragraph 709.

Chapter 3

—◆—

Enantiodromia: From Roman Catholicism to Newtonian Science

The transition from Roman Catholicism to Newtonian Science is one instance of enantiodromia. This enantiodromia generated a counter movement away from Catholicism, which was dominated by religious doctrine and skepticism about science. I describe this instance of enantiodromia by explaining the extreme conscious focus on the practice of selling indulgences and other behaviors that characterized Roman Catholicism in the sixteenth century. Then, I describe the emergence of counter movements in opposition to the extreme conscious attention being devoted to Catholic religious practices. The counter movement was triggered by a number of bifurcations. The bifurcations that I choose to highlight this enantiodromia are the Protestant Reformation and the establishment of a Jesuit Order. The Reformation provided a trigger by reducing the size and influence of the Catholic Church. The establishment of a Jesuit Order was another trigger, in the sense that the Jesuits were instrumental in training scientists, whose work eventually shifted authority from religion to science. There were other bifurcations. I select these two because of their compelling influence in the paradigm shift from Roman Catholicism to Newtonian Science.

Roman Catholicism

In the Renaissance culture, The Roman Catholic Church was a patron of expensive art. That was when art patronage was a symbol of power, and before the separation of church and state. When the separation of church and state got underway, money that had been available to the Church moved to the nation-states in the form of taxes. The diminished budget available for art collection and patronage reduced the power of the Catholic Church in Western societies.[1] The church searched for new sources of money for art that depicts Catholic doctrine.

To raise money for art, the Catholic Church began selling indulgences. According to the Catholic.com web site: "An indulgence is a remission before God of the temporal punishment due to sins whose guilt has already been forgiven." [2] Indulgences do not forgive sins. They address punishments after sins have been forgiven. There were questions about indulgences. Can a person obtain an indulgence only for oneself? Can a person obtain an indulgence for another person who is dead and in Purgatory? When obtaining an indulgence for another, must one first confess one's own sins? Uncertainty surrounding the penance related to indulgences threatened to sever the connection between the confession of sin and the achievement of salvation.[3] A turning point occurred in the early sixteenth century, when a German friar named Johann Tetzel was found selling indulgences accompanied by the jingle:

"When a penny in the coffer rings,
A soul from Purgatory springs." [4]

Tetzel's approach increased concerns about the sale of indulgences, but it is German friar Martin Luther who is credited with initiating the Protestant Reformation.

Bifurcations: Protestant Reformation, Jesuit Order

The Protestant Revolution brought about a significant bifurcation of the Roman Catholic Church. A bifurcation is a split of something into two parts. Historian Richard Tarnas explains a cause of the Protestant Reformation, which split the Roman Catholic Church into Catholics and Protestants:

"The proximate cause of the Reformation was the papacy's attempt to finance the architectural and artistic glories of the High Renaissance by the theologically dubious means of selling spiritual indulgences. (Johann) Tetzel, the travelling friar whose sale of indulgences in Germany provoked Luther in 1517 to post his Ninety-five Theses, had been so authorized by the Medici Pope Leo X to raise money for building Saint Peter's Basilica. An indulgence was the remission of punishment for a sin after guilt had been sacramentally forgiven – a Church practice influenced by the pre-Christian Germanic custom of commuting the physical penalty for a crime to a monetary payment. To grant an indulgence, the Church drew from the treasury of merits accumulated by the good works of the saints, and in return the recipient made a contribution to the Church. A voluntary and popular arrangement, the practice allowed the Church to raise money for financing crusades and building cathedrals and hospitals. At first applied only to penalties imposed by the Church in this life, by Luther's time indulgences were being granted to remit penalties imposed by God in the afterlife, including immediate release from purgatory. With indulgences effecting even the remission of sins, the sacrament of penance was itself seemingly compromised." [5]

In 1517, friar Martin Luther initiated a bifurcation in the Catholic Church by publishing his Ninety-five Theses, a list of practices that were counter to the Catholic faith. These Theses challenged the Catholic Church. By 1520, Luther had composed a succinct theological message that was liberating and easy to understand: salvation is free; indulgences are not necessary. [6] Luther's work diminished the laity's confidence in Catholic leadership and created a schism in the Church. The schism led to the Protestant Reformation. One of the effects of the Reformation was a secularizing effect on Western culture. [7] Reformers eliminated dependence on the clergy to interpret scripture. Individuals were encouraged to interpret the scriptures for themselves. By shifting theological authority from the clergy to the laity, the Reformers individualized ethics and morality. That led to the development of secularization. The Protestant religion became fractured into multiple religions. People choose their beliefs. There was no centralized institution guiding people with a common set of theological rules. Religion lost its authority. Authority shifted from the Catholic Church to individuals and government institutions.

Luther reduced the seven Catholic sacraments to three sacraments for Protestants: Baptism, Confirmation and Confession. The explanation for the reduction was that these three sacraments are based on Biblical theology, while the other sacraments are based on Scholastic theology. Reformers interpreted the Bible in a literal way, while regarding Scholastic theology as too abstract. Biblical theology is literal. It aims to find out what the authors of the Bible regarded as divine guidance for the conduct of human life, at the time when the Bible was written. Scholastic theology is abstract. It is an intellectual interpretation of divine guidance for humans. It was developed in the Western world and used by the Catholic Church. By dispensing with the sacramental authority of the Catholic Church, Luther enabled the Reformers to become more self-reliant. By separating Biblical theology from Scholastic theology, Reformers increased the distinction between God's will and human intelligence.[8] This distinction "allowed the modern mind to approach the world with a new sense of nature's purely mundane character, with its own ordering principles that might not directly correspond to man's logical assumptions about God's divine government." [9] The reformers' limiting of the human mind to a this-worldly knowledge was precisely the prerequisite for the opening up of that knowledge." [10] By separating God's divinity from human knowledge, the Protestant Reformation created a path for openness to Newtonian Science.

A second bifurcation was created by the establishment of a Jesuit Order as a reaction to the effect of the Protestant Reformation. Large numbers of the faithful were leaving the Catholic Church to join Protestant Churches. The Jesuit Order was a Counter-Reformation movement. It was established as part of an effort to restore confidence in the Catholic Church. The initiator of the Jesuit Order was Ignatius de Loyola, a Spanish priest. He and six young men who had met at the University of Paris created a plan for the new order. In 1540, Pope Paul III approved Loyola's plan for the creation of the Jesuit Order. [11] Education and scholarship were the priorities of the Jesuit Order. Loyola and his team established educational institutions and programs in Europe, Africa, Asia and the Americas. The Jesuit Order was the first religious order to operate universities and colleges as a separate ministry. They taught Renaissance humanism, classical literature, philosophy, languages, arts and sciences. When Loyola died in 1556, he had established thirty-three educational institutions across South America, Asia, Africa and Europe. [12] By teaching science, the Jesuits began to create generations of professionals who had the skills and knowledge to be mathematicians, astronomers and physicists. These are the professionals who led the way to Newtonian Science.

Newtonian Science

Newtonian Science was a counter movement to the Roman Catholicism. Newtonian Science encompasses the scientific work accomplished by Isaac Newton (1642 – 1727). He was born in England and he was a Protestant. He belonged to the Anglican Church which made a point of separating its own practices from those of the Catholic

14

Church. Newton's approach to science was that by observations and experiments, physicists can apply empiricism to generate laws about nature. In 1666, his achievements included the proposal that gravitational force holds the moon in place, the invention of calculus and the demonstration that white light is a mixture of colored lights. [13] In 1687, his book "Principia Mathematica" was published. This publication included the three laws of motion which became the foundation of classical mechanics. It also included the law of gravitational forces.

Newton began his work where his predecessors had ended theirs. In 1543, Copernicus published his work on heliocentricity, which puts the sun at the center of the universe with the planets moving around it. Galileo Galilei promoted Copernicus' work on heliocentricity and built telescopes for people to view planets orbiting the sun. Newton was born on the day in March 1727 when Galilei died. Newton would eventually carry on Galilei's work on the mathematical science of motion and take it to completion. [14] Here are Newton's three laws of mechanics describing the motion of a body: [15]

"The first law states that a body remains at rest or in uniform motion in a straight line unless acted upon by a force.

The second law states that a body's rate of change of momentum is proportional to the force causing it.

The third law states that when a force acts on a body due to another body, then an equal and opposite force acts simultaneously on that body."

Copernicus and Galilei did not settle the question of what held the planets in place. German scientist Johannes Kepler (1571 – 1630) had composed laws of motion by trial and error, but did not have a theoretical model for gravity. [16] Newton built on Kepler's work to develop a law of gravity defined as follows: [17]

15

"Every particle of matter in the universe attracts every other particle with a force that is directly proportional to the product of the masses of the particles and inversely proportional to the square of the distance between them."

Although scientists made significant progress in classical mechanics, the Catholic Church was rather slow to embrace science. Copernicus published his work on heliocentricity in the year he died, 1543. The Catholic Church put his book about heliocentricity in their Index of Forbidden Books, where it remained for about 200 years. In 1633, the Catholic Church found Galilei guilty of heresy for promoting heliocentricity. He was sentenced to life imprisonment but was allowed to spend the rest of his life under house arrest. [18]

The Jesuits made spectacular progress in educating young people about the sciences, among other subjects. What is noteworthy about the transition from religion to science is that the Catholic Church, which was opposed to science, ended up producing Jesuit-trained scientists like Copernicus and Galilei, who discovered and promoted the science of heliocentricity. Tarnas points out that it was no accident that Galilei, Descartes, Voltaire and Diderot all received Jesuit education. [19]

This chapter demonstrates enantiodromia in the transition from religion to science. When the Western society's conscious focus on religion reached an extreme, there were two bifurcations that set society on a path to science. The Roman Catholic Church was split into a Catholic congregation and a Protestant congregation. The Jesuit Order split education into faith-based education and science-based education. These bifurcations paved the way for the emergence of Newtonian Science from the unconscious psyche. This transition from religion to science shows the interplay of opposites between consciousness and unconsciousness, which is the hallmark of enantiodromia.

NOTES

1. Link to information about Catholic art patronage: https://www.artsy.net/article/artsy-editorial-happened-catholic-churchs-art-patronage.
2. See details about the nature of indulgences: https://www.catholic.com/tract/myths-about-indulgences.
3. Link to uncertainties about indulgences: https://www.britannica.com/topic/indulgence.
4. Link to Johann Tetzel's jingle that accompanied his sale of indulgences: https://www.britannica.com/topic/indulgence.
5. See Richard Tarnas' book "The Passion of the Western Mind" pp 233 – 234.
6. See Luther's theological message: https://www.britannica.com/topic/indulgence.
7. See Tarnas' book, p 240.
8. See Tarnas' book, p 241.
9. See Tarnas' book, p 241.
10. See Tarnas' book, p 241.
11. Link to information about the establishment of the Jesuit Order: https://www.history.com/this-day-in-history/jesuit-order-established.
12. See the continents where Loyola established schools by the time he died: httpsl://www.encyclopedia.com/people/philosophy-and-religion/saints/saint-ignatius-loyola.
13. See Newton's achievements in 1666: https://www.britannica.com/biography/Isaac-Newton.
14. See information about Newton having carried on the work of Galilei in the laws on the science of motion: https://www.britannica.com/biography/Isaac-Newton.
15. See definitions of Newton's three laws of motion: https://www.dictionary.com/browse/newton-s-laws-of-motion.
16. See information about Kepler's work on gravity: https://www.thoughtco.com/newtons-law-of-gravity-2698878.

17. See definition of Newton's law of gravity:
    https://www.thoughtco.com/newtons-law-of-gravity-2698878.

18. See information about Galilei's house arrest:
    https://www.history.com/this-day-in-history/astronomer-galileo-dies-in-italy.

19. See Tarnas' comment on the Jesuit education of scientists: "The Passion of the Western mind", p 247.

Chapter 4

―∾―

Enantiodromia: From Enlightenment to Romanticism

To explain this instance of enantiodromia, I begin by describing the extreme focus on science and natural laws that characterized the Age of Enlightenment. While there may be other bifurcations, I chose the 1755 Lisbon Earthquake as a major bifurcation that created a split from Enlightenment because of the enormous impact it had on multiple countries. The fact that science was inadequate to explain an earthquake that killed so many people and destroyed so many buildings shook people's confidence in science. Then, I describe the emergence of a counter movement to the extreme conscious attention being devoted to science. The counter movement was Romanticism. In the Age of Romanticism, the world became an organic whole in which people could indulge their artistic sensibilities.

Enlightenment

The Age of Enlightenment was a movement that occurred between the seventeenth and eighteenth centuries, when science was celebrated as the means by which humans learned to understand the world and used that knowledge to improve the conditions of their lives.[1] Enlightenment established science as humanity's highest mental achievement. The works of Roger Bacon, Nicolaus Copernicus, Galileo Galilei, Johannes Kepler and others were generating the modern scientific methods and laws that controlled nature. The Enlightenment was characterized by an extreme adherence to the notion that the world is governed by natural scientific laws which guide human affairs. The implication was that if we could discover enough of these natural laws, we could create orderly and prosperous societies.

Bifurcation: The 1755 Lisbon Earthquake

According to the principle of enantiodromia, the Age of Enlightenment was poised for a change because it had become the extreme focus of conscious attention. One event that was pivotal in bringing about the change was an earthquake. It occurred in Lisbon, the capital city of Portugal on November 1st of 1755. On that day, three earthquake tremors struck killing an estimated 50,000 people. [2] It was All Saints Day and Catholics were in church to celebrate. Many of them died when their churches collapsed. As the earthquake rocked the churches, candles tumbled igniting flammable materials and creating fires that burned in Lisbon for five days after the earthquake.[3] Later that month, there were significant aftershocks and tsunami waves. The disasters of November shattered the sense of wellbeing of the people living in the area. The epicenter of the earthquake was in Lisbon, but it was felt in several European countries and as far away as Morocco. The European communities searched for an explanation of the devastation. The science of the Enlightenment provided no answers. The earthquake had not been predicted by any natural law. Nor was there any logical explanation of why the earthquake occurred. It became clear that the natural laws of science did not offer any assurance that there will be order in nature, or any protection from natural disaster. Disillusioned with the silence of science on the reason for the earthquake, people sought a religious explanation. Do disasters happen at God's will? Was the devastation of the Lisbon earthquake God's punishment for human sins? The widespread and unrelenting ideas about the natural laws had made the Enlightenment an extreme in the conscious attention. Now, society was ready for a change. The failure of the natural laws of the Enlightenment to explain the disasters (earthquake, fires, tsunami, flooding) of nature was a turning point. The vacuum left by disillusionment with Enlightenment made room for an emergence from society's unconsciousness. That emergence turned out to be Romanticism.

Romanticism

The Age of Enlightenment had brought all of nature under the light of scientific investigation. It was followed by the Age of Romanticism which began in the late eighteenth century. By the nineteenth century, Sigmund Freud had extended the investigation of nature to include the psyche. Ironically, Freud's investigation revealed that there were non-rational forces underpinning the rational, scientific certainties of Enlightenment. So, the interpretation was that the human being is conditioned by non-rational forces, rather than purely rational forces. [4]

The Age of Romanticism embraced the notion that humans do not always behave rationally. Philosopher Immanuel Kant was an influential figure in the transition from Enlightenment to Romanticism. He downplayed the idea that the mind plays a passive role in understanding the structure of the world through natural laws. He promoted the idea that the world order is actively structured by the mind. He believed that what constitutes science is what the mind has made available to science. Historian Richard Tarnas endorsed Kant's idea by stating that human knowledge does not conform to objects; objects conform to human knowledge. [5] That outlook did not bring science to an end; science continues to flourish in the twenty first century. What the Kantian outlook accomplished was the provision of a stepping-stone for the emergence of a counter movement in opposition to the extreme conscious attention being devoted to science and natural laws. The counter movement was Romanticism. In the Age of Romanticism, humans began to regard the world as an organic whole canvas for them to exercise their artistic abilities. The new-found outlook was reflected in domains where the human inner world prevailed, for example, literature, paintings and psychology. The psychology of Freud redefined humankind in terms of a conscious component capable of logical thought, and an unconscious component capable of non-logical mentation, including imagination and creativity. Tarnas sums up the agenda of Romanticism: explore the mysteries of the interiority, plumb the depth of the soul, and bring the unconscious into consciousness. [6]

This chapter demonstrates enantiodromia in the transition from Enlightenment to Romanticism. At the point when the society's conscious focus on natural laws reached an extreme, there occurred a bifurcation that set society on a path to a counter movement. The earthquake was instrumental in splitting society's attention between a reliance on the laws of nature and an exploration of the interiority of the psyche. The laws of nature continued to be researched and applied, but they no longer dominated the society's conscious attention. This transition from Enlightenment to Romanticism demonstrates the interplay of opposites between consciousness and unconsciousness that is characteristic of enantiodromia.

NOTES

1. See information about the characteristics of the Age of Enlightenment:
   (https://www.britannica.com/event/Enlightenment-European-history).
2. Link to information about the 1755 Lisbon earthquake:
   (https://www.history.com/this-day-in-history/earthquake-takes-heavy-toll-on-lisbon).
3. See information about the fires that followed the Lisbon earthquakes: (https://lisbonlisboaportugal.com/Lisbon-information/1755-lisbon-earthquake.html).
4. Link to information about the non-rational underpinnings of science in Richard Tarnas' book "The Passions of the Western Mind", p 329.
5. See Tarnas' comment on Kant's idea, p 346.
6. See Tarnas' comment about the agenda of Romanticism, p 368.

Chapter 5

―――∾――――

Enantiodromia: From Modernism to Postmodernism

To describe this instance of enantiodromia, I first explain the extreme focus on progress and objectivity that characterized Modernism. Then, I present two bifurcations from Modernism. One was World War II; the other was technology. I follow the bifurcations with a description of the nihilism and subjectivity that have been the hallmarks of Postmodernism. This example of enantiodromia covered a period from the late twentieth century to the twenty first century. Although there may be other bifurcations, I selected World War II and technology because of the enormity of the effects on multiple nations and the major swing in outlook to the opposite of Modernism.

Modernism

Modernism was a cultural movement in the period from the late nineteenth century through to the early twentieth century. The movement encompassed a wide variety of cultural aspects.

Traditional forms of art, architecture, history, music, law, philosophy, politics and religion were considered inappropriate by people in the new industrialized world. The desire for progress engendered innovations like Pablo Picasso's abstract art; Josef Matthias

Hauer' twelve tone musical composition style; and Frank Lloyd Wright's "Falling Water" architectural design style.

Emerging from the rebellious outlook that characterized the early twentieth century, Modernism was an approach that sought to improve modern civilization, prompted by a rejection of the European culture which had become corrupt and complacent due to being bound by the artificialities of a society that was preoccupied with appearances.[1]

The Modernists were dissatisfied with what they perceived as the moral bankruptcy of everything European and that led them to explore other alternatives, with the intention of undermining tradition and authority, while hoping to transform contemporary society.[2] Modernism had been shaped by industrial societies. Although Modernists recognized the benefits of objectivity and socio-technological progress, they rejected belief systems of institutions related to religion, art, architecture and politics. They saw belief systems as restraining the human spirit. The rules and regulations of institutions did not reflect life in the way that Modernists experienced life. The Modernists sought to discover the underpinnings of institutional rules and liberate society from the limitations imposed by institutions. In spite of their skepticism about institutions and bureaucracy, the Modernists were more comfortable living in cities than in nature. Modernism achieved the most progress in those societies that were dominated by consumerism and capitalism.

There was a theme of self-consciousness that pervaded the innovations of Modernism. It found its way into the arts as well as the sciences. The self-consciousness was a progressive trend of thought that affirmed the power of human beings to create, improve and reshape their environment with the aid of experimentation, scientific knowledge and technology. The relatively new discipline of psychology turned the analysis of human experience inward and inspired a more abstract outlook.[3] Modernism sought to revise all areas of life, with the intention of improving the progress of humanity. World War II created

24

the first bifurcation in Modernism. The second bifurcation was the enormous technological growth that followed World War II.

Bifurcation: World War II, Technology

The disillusionment that followed World War II (1939 – 1945) created a significant backlash to Modernism.[4] Modernism was a culture that embodied ideals about the progress of society based on what was valued at the time. World War II presented a major bifurcation for Modernism. Although war itself was not among the ideals of Modernism, it had been possible to fit World War II into the Modernist view of progress. War had found a place in the context of Modernism. War was also consistent with the competition inherent among nation states. The destructiveness of the war had found a place in outlook upon which Modernism had been established.

Paul Crosthwaite is a lecturer in English Literature and member of the Centre for Critical and Cultural Theory at Cardiff University in the United Kingdom.[5] He published "Trauma, Postmodernism, and the Aftermath of World War II" in 2009. Here is an excerpt from Crosthwaite's summary of the Postmodern view in the aftermath of World War II: [6]

"... [T]he horrors of World War II are viewed as arising from, and rendering all too apparent, the inherent antagonistic character of the nation state; the dominance of instrumental rationality within modern bureaucratic, administrative, and logistical systems, and the consequent adaptability of such systems to mass reification, and ultimate extermination of human life; the capacity of political and military institutions to dominate, discipline, and control, the natural environment and its human inhabitants; the propensity of scientific and technological expertise to be channeled into the production of highly efficient technologies of war and genocide; the co-construction of capitalism, imperialism and militarism; the liability of 'western' culture to objectify, oppress, and eliminate its ethnic and racial 'others'; and the bloody futility of others to attempt to make over the world in the image of some grand, Utopian vision, whether of the right or the left."

The death toll of World War II is estimated as being between 50 million and 80 million.[7] On-the-ground combat in various countries, the dropping of bombs and the Holocaust all combined to produce the numbing effect of an existential nihilism on civilization. Life had lost its meaning and its purpose. One effect of World War II was that it had a stimulating impact on technology. Shortly after the war, there were inventions of computer, TV and Internet. This upsurge in technological innovation created another bifurcation of Modernism. These innovations enhanced the lifestyles of many societies. However, the rapid change and the transition to a digital world created a disorientation for many who were not prepared for a virtual lifestyle. Societies were further disillusioned by loss of jobs to technology and perplexed by the personas of social media.

Postmodernism

Postmodernism was a reaction to Modernism. While Postmodernism opposed the basic tenets of Modernism, it had few tenets of its own. Postmodernism began in the late twentieth century, following World War II, and continues into the twenty first century.[8] Words commonly associated with Postmodernism are nihilism and subjectivity. Jacques Derrida, who is credited with being the founder of Postmodernism, used the French word *"différance"* to characterize the rather circular Postmodernist position on meaning: [9]

"[A]ny given word or set of meanings derives from an imprecise definition in terms of other meanings, which are *themselves* imprecise."

The Psychology Wiki offers a less circular comment:

"Postmodernism is a culture that lacks a clear central hierarchy or organizing principle, while it embodies complexity, contradiction, ambiguity, diversity and interconnectedness." [10]

Two of the main historical events that catalyzed the emergence of Postmodernism are bifurcations that created splits in Modernism. World War II split the world into two collections of nations with opposing world views. The Allied Powers (Britain, France, the United States and the Soviet Union) adopted a political outlook of democracy, while the Axis Powers (Germany, Italy and Japan) favored communism. The worldwide military conflict of World War II was the deadliest war in history. It had the effect of reducing the extreme focus of society's mind on Modernism. The destruction of cities in multiple countries took conscious attention away from Modernism. When reflecting on the society and human nature, people lost confidence in their own judgment, resulting in a sense of loss, insecurity and anxiety. [11] After World War II, the hope of a peaceful world had collapsed, while the reliability of human rational thought and self-control were seriously impaired.

World War II was instrumental in the development of new technology, which created a second bifurcation. From 1950 onward, there were computers, televisions and access to devices that split lifestyles into the virtual lifestyle and the reality lifestyle.

Every new technological invention triggered a significant change in people's lifestyles. In one sense, technology enhanced human lifestyles by enabling people to be smarter and more efficient. In another sense, technology separated people from reality by establishing virtual lifestyles. What used to be a clear sense of an individual's identity, gave way to a disruption of the sense of "self". [12] In the twenty first century, people continue their struggle to balance the benefits of technology with their disrupted sense of self.

This chapter demonstrates enantiodromia in the transition from Modernism to Postmodernism. The conscious part of the psyche was focused on extreme attention given to improving modern civilization.

World War II was the first major event that shattered the relatively peaceful world and split it into two opposing forces of democracy and communism. Technology was the second bifurcation. It had the effect of splitting lifestyles into those that are real and those that are virtual. The bifurcations shifted the world's attention from Modernism to a Postmodernism that led people to question the purpose of life and their sense of identity. The transition from Modernism to Postmodernism is an example of the interplay of opposites between consciousness and unconsciousness. That interplay is characteristic of enantiodromia.

NOTES

1. See information about the characteristics of Modernism: https://www.mdc.edu/wolfson/Academic/ArtsLetters/art_philosophy/Humanities/history_of_modernism.htm.
2. See comments about contemporary society in the Age of Modernism: (https://www.mdc.edu/wolfson/Academic/ArtsLetters/art_philosophy/Humanities/history_of_modernism.htm).
3. See comments about psychology in the era of Modernism: (https://www.history.com/topics/art-history/history-of-modernism-and-post-modernism).
4. See comments on the effect of World War II on Modernism: (https://psychology.wikia.org/wiki/Postmodernism)
5. See Paul Croswaithe's comments about World War II in his book "Trauma, Postmodernism, and the Aftermath of World War II": (https://www.research.ed.ac.uk/portal/pcrosthw).
6. See excerpt from Paul Crossthwaite's comments about World War II: (https://www.palgrave.com/us/book/9780230202955
7. See statistics on the estimated death toll of World War II: https://www.nationalww2museum.org/students-teachers/student-resources/research-starters/research-starters-worldwide-deaths-world-war.

8. See comments about the beginning of the era of Postmodernism: (https://psychology.wikia.org/wiki/Postmodernism).

9. See Jacques Derrida's definition of "Difference": (https://rationalwiki.org/wiki/Postmodernism).

10. See Psychology Wiki's comments on the nature of a Postmodern culture: (https://psychology.wikia.org/wiki/Postmodernism).

11. See comments about the aftermath of World War II: (https://avantgarde-jing.blogspot.com/2010/03/emergence-characteristics-and.html).

12. See comments about a changing sense of identity in the era of Postmodernism: (https://avantgarde-jing.blogspot.com/2010/03/emergence-characteristics-and.html).

Chapter 6

———— ∿ ————

Enantiodromia: From Empiricism to Unconscious Dynamics

For this instance of enantiodromia, I describe the transition from extreme conscious focus on empiricism to a new approach that focuses on unconscious dynamics. Empiricism is a style of obtaining knowledge by relying on observable facts to produce decisions that are not questionable. I use the expression "unconscious dynamics" to refer to the works of Immanuel Kant, Niels Bohr, Sigmund Freud and Carl Jung, all of whom pointed to ways of obtaining knowledge that are not observable to the naked eye. They obtained knowledge by unobserved speculation, intuition, deduction and inference. I point out that unconscious dynamics compromise the factual observations of empiricism. This example of enantiodromia covers a period from the late eighteenth century to the middle of the twentieth century, and the transition involved multiple bifurcations. I selected bifurcations from the works of German philosopher Immanuel Kant, Danish physicist Niels Bohr, Austrian neurologist Sigmund Freud and Swiss psychologist Carl Jung.

Empiricism

In the age of the empiricists, knowledge was obtained by a subject who observed nature with an objective approach. The assumption was that scientists could conduct experiments based on the collection of factual data and by applying empirical methods. The empiricists were confident that knowledge could be achieved exclusively by empirical methods. Their confidence experienced serial bifurcations by Immanuel Kant, Niels Bohr, Sigmund Freud and Carl Jung.

Bifurcations: Kant, Bohr, Freud & Jung

The works of Kant, Bohr, Freud and Jung all initiated bifurcations, which together created a shift away from the empirical paradigm, that had an extreme focus on observable facts. The bifurcations generated new outlooks that blurred our understanding of just how does our psyche process information. Changing outlooks in the domains of philosophy, physics, neurology and psychology all combined to produce a swirl of uncertainty about how humans acquired knowledge. What emerged from the swirl was a new paradigm which made us aware that our striving for empiricism was actually constrained by influences arising from the interior world of our psyche.

Kant's Bifurcation of Empirical Observation:
Sensory Input & A Priori Structures

In the late eighteenth century, Immanuel Kant created a bifurcation in empiricism by proposing two types of observations. One type of observation is obtained through sensory input. Sensory observations are those that we take in through our five senses, for example, what we see, hear and measure. The other type of observation comes about through a priori structures of our cognition. Examples of a priori structures embedded in our cognition are space and time. Since time and space do not exist in nature independently of cognition, Kant considered them to be among the ordering structures of the mind. He proposed that our efforts at empiricism are compromised by the structures of our cognition.

Kant was born in Germany and lived in the period from 1724 to 1804. His education at the University of Konigsberg made him aware of the science of Newtonian physics and the beliefs of Judeo-Christian religions. In the late eighteenth century, Kant wrote and published three Critiques in which he explained his view that the basic underpinning of human understanding is autonomy. [1] The Critiques are: the "Critique of Pure Reason" (1787), the "Critique of Practical Reason" (1788) and the "Critique of the Power of Judgement" (1790). These Critiques proposed that everything we know is constrained by the way our mind works. Kant questioned the absolute certainty of empiricists who claimed that we can know nature with certainty by applying empirical methods. He also questioned the religious belief of those who claimed certainty based on the gift of divine providence.

As a philosopher, Kant brought a fresh perspective to the topic of how humans think. He brought together the scientific outlook and the religious outlook to show us that both outlooks depend on authorities. Science depends on the authority of scientists such as Newton, who established laws of physics. Religion depends on the authority of the clergy to interpret religious doctrines. At the same time, Kant proposed that our outlook on both science and religion actually depend on how we think. In a significant departure from philosophy of the time, Kant showed us that we understand both science and religion based on an autonomy that derives from our internal cognitive structures. Kant's bifurcation of empirical observation alerted us to the realization that, when we apply empirical methods, our outward-facing attempt at empiricism is shaped by the autonomy of our internal cognitive structures.

Bohr's Bifurcation of Physics: Classical Physics & Quantum Physics

In the early twentieth century, physics experienced a bifurcation into classical physics and quantum physics. Classical physics is made up of laws that facilitate empiricism because they apply to large things that can be observed and defined with specificity. Quantum physics is a collection of principles that require inference because they apply to small things whose definitions involve probability. Classical physics describes matter at the macro level, such as Newton's laws of motion. Quantum physics is about particles at the subatomic level, for example, the electrons of an atom, which is not visible to the naked eye. In keeping with the new physics, Niels Bohr made us aware that the certainty with which we pursue empiricism in classical Newtonian physics does not apply in quantum physics, where inference is required. That is because matter at the atomic level has both wave properties and particle properties. An electron can behave as a wave function or a particle, depending on how it is measured. The measurement of a wave is expressed as a probability. The measurement of a particle is expressed as discrete properties, such as position and speed.

Niels Bohr (1885 – 1962) was born in Denmark and considered a prominent scientist of the twentieth century. He studied at the University of Copenhagen, where he earned a doctorate degree in physics.[1] In 1922, Bohr received the Noble Prize in physics for his work on the atomic structure and quantum physics.[2] Bohr's model of the atomic structure included positively charged photons clustered as a nucleus at the center, with negatively charged electrons circling the nucleus in orbits. The electrons can jump from one orbit to another. A jump involves a change in energy level for the electron. Bohr also defined the Complementarity Principle and demonstrated that at the micro level (atomic and subatomic scale), complete knowledge of matter requires knowledge of the wave properties as well as particle properties. The Complementarity Principle applies to quantum phenomena, such as light waves, photons and electrons. Quantum phenomena will behave like a wave if measured with one type of instrument, but behave like a particle if measured with a different type of instrument. Since the choice of one measuring instrument eliminates other possible instruments, the observer has an effect on the measurement, by selection of instrument. Bohr's contribution to the bifurcation of physics shows that the empiricism of classical physics does not apply to quantum physics, which involves non-empirical topics such as probabilities and speculative hypotheses.

Freud's Bifurcation of Psyche: Consciousness & Unconsciousness

Sigmund Freud (1856 – 1939) created a bifurcation by separating the human psyche into conscious and unconscious portions. While pursuing a career in neurology, he compared available treatments. When he noticed that certain illnesses had no origin in the physical body, he created a new discipline which he called "Psychoanalysis" to denote the practice of treating illness that originates in the mind. [4]

Freud was born in Austria and earned his qualification as doctor of medicine at the University of Vienna. He chose neurology as his area of specialty. Toward the end of the nineteenth century, Freud proposed a psychoanalytic theory that characterized the psyche as having two ways of processing information: the conscious part of the psyche processes information in a rational manner, while the unconscious part of the psyche processes information in a non-rational manner. [5] Freud published "Interpretation of Dreams" (1901) which explained that dreams are the language of the unconscious.[6] Freud's Psychoanalytic Theory proposed a structural model of the human psyche that includes conscious and unconscious dynamics.[7] The conscious portion of the psyche accommodates the kind of thinking that supports empiricism. The unconscious portion of the psyche communicates through non-empirical means such as dreams, emotions and imagination.

The empiricism that could be applied to physical disorders could not be applied to disorders of the psyche. Psychoanalysis diminished the empiricism of medical treatment by introducing the notion that there is an unconscious aspect of our psyche, which does not submit to the empirical methods that had been the basis for medical treatment. The psychoanalytic theory posits that humans possess a psyche made up of two main components that process information differently. The conscious portion processes sensory observations and logical thoughts, while the unconscious portion processes imagination, dreams and fantasies. He further explained that we can access consciousness at will, but we cannot access unconsciousness directly. Freud's bifurcation of the human psyche informed us that the empiricism which applied to the body was not adequate for understanding the mind. The health of our psyche involves an integration of both conscious content and unconscious content. That integration requires both empiricism and unconscious dynamics.

Jung's Bifurcation of Unconsciousness: Personal & Collective Unconsciousness

Early in the twentieth century, Carl Jung created a bifurcation of Freud's unconscious portion of the psyche. Jung split Freud's unconscious into a personal unconscious and a collective unconscious. He also rendered empiricism subservient to our unconscious dynamics. In Jung's model of the psyche, consciousness is where we strive for empiricism that includes observation and logical thought. The personal unconscious is a storage of personal experience that was once conscious, but was repressed. The collective unconscious is a universal reservoir of ancestral experience that is common to all of humankind. The conscious part of our psyche is influenced by the unconscious parts, even when we are not aware of it. When we are not aware of the influence, our empirical efforts are compromised. When we are aware of the influence, we can integrate our conscious and unconscious capabilities to our advantage.

Carl Jung (1875 – 1961) defined a new discipline called "Analytical Psychology" based on the notion that the psyche is a self-regulating system in which the ego mediates interaction between consciousness and the heritage-laden archetypes that populate unconsciousness. [8] Archetypes are structures of the psyche that constitute pre-dispositions for people to think, feel, perceive and act in specific ways. [9] Although Jung's "Analytical Psychology" was not a proven theorem, it reduced our extreme reliance on empiricism by showing that our information processing is compromised by the influence of archetypes in the unconscious part of our psyche. Jung's bifurcation of unconsciousness eroded our confidence in our understanding of our psyche. The discovery of archetypes makes it necessary to understand our interactions with universal ancestral heritage.

Unconscious Dynamics

Together, the works of Kant, Bohr, Freud and Jung demonstrate enantiodromia which began with an extreme culture of empiricism that focused on what is observable and measurable. That focus shifted toward a respect for other capabilities of the psyche, which includes intellectual speculation, hypotheses and intuitions about matters not directly observable or measurable. Kant, Bohr, Freud and Jung brought it to our attention that invisible though it may be, our psyche does play a significant role in our empirical efforts. This enantiodromia is made up of a series of bifurcations in the domains of philosophy, physics, neurology and psychology. Kant, Bohr, Freud and Jung all created bifurcations that demonstrate a progressive shift from the sovereignty of empiricism toward a discovery of unconscious dynamics. With each bifurcation, Kant, Bohr, Freud and Jung showed us that there is more to the psyche than empiricism could discern. Each bifurcation taught us something new about the unconscious dynamics of the human psyche.

This chapter provides an example of enantiodromia in the transition from empiricism, which is outward facing to the unconscious dynamics, which are inward facing. During the era of empiricism, the conscious part of the psyche was focused on extreme attention given to the external world. Through a series of bifurcations, Kant, Bohr, Freud and Jung split empiricism repeatedly. The effect of the serial splitting was that our attention shifted from observing and measuring the external world to discovery of the interior world of the psyche. The transition from empiricism to unconscious dynamics is an example of the interplay of opposites between consciousness and unconsciousness that illustrates enantiodromia. This enantiodromia transported us on an evolutionary journey from an extreme reliance on empiricism, to a nuanced appreciation of the capabilities of our psyche, and later to an awareness that there are unconscious dynamics at play in our psyche.

On that journey we discovered that our psyche has functionality previously unknown to us. In addition, we saw opportunities for a more explicit knowledge of our psyche than existed in past centuries.

NOTES

1. See Kant's view of autonomy: https://plato.stanford.edu/entries/kant/.
2. See information about Neils Bohr's education: https://www.britannica.com/biography/Niels-Bohr.
3. See information about Neils Bohr's Nobel prize: https://www.livescience.com/32016-niels-bohr-atomic-theory.html.
4. See information about Sigmund Freud's development of psychoanalysis: (https://www.notablebiographies.com/Fi-Gi/Freud-Sigmund.html).
5. See Freud's separation of consciousness from unconsciousness: (https://www.britannica.com/biography/Sigmund-Freud).
6. See information about Freud's publication of "Interpretation of Dreams": (https://www.livescience.com/54723-sigmund-freud-biography.html).
7. See Freud's conscious and unconscious dynamics: (https://www.enotes.com/research-starters/freuds-structural-model-psyche).
8. See information about Carl Jung's Analytical Psychology: http://journalpsyche.org/tag/personal-unconscious-and-collective-unconscious/.
9. See information about archetypes: (https://academyofideas.com/2016/01/introduction-to-carl-jung-the-psyche-archetypes-and-the-collective-unconscious/).

Chapter 7

————— ∿ —————

Nature of the Psyche

Proponents of Singularity wonder what exactly is consciousness and where might it be located. If technology is going to outsmart humans, technology will need some functionality that is equivalent to consciousness. Psychologist Carl Jung defined the psyche as: The totality of all psychic processes, conscious as well as unconscious. [1] Just as every human has a body, every human has a psyche. The body is the tangible part of us. The psyche is the intangible part. The tangible and intangible parts work together. Both parts interact as we go through the activities of daily living. They are engaged when we interact as individuals, when we sit alone in introspection and when we function collectively as a society. Though illusive and invisible, the human psyche makes its presence known by its influence. It influences the way we engage the external environment, our relationships with other people and the way we manage our internal world. The psyche has resources to perform activities that involve knowledge, memory, information processing, language skills and computation capabilities.

Carl Jung described the structure of the psyche in "Structure & Dynamics of the Psyche".[2] That is my source for describing the components of the psyche that I select for this study. The psyche has two main areas of functioning: the conscious realm and the unconscious realm. In the conscious realm, the psyche processes information. The processing is time-specific, data-specific, text-specific and heavily reliant on explicit knowledge. Explicit knowledge is direct knowledge that we are able to articulate and transfer to others. In the unconscious realm, the psyche processes images, fantasies, ideas, symbols and dreams. The processing depends on extraction of meaning from semantic structures, symbols and images, while being heavily reliant on implicit knowledge. Implicit knowledge refers to indirect knowledge that is abstract and not readily transferable because it is not procedural. It is knowledge acquired independently of a person's conscious awareness. The psyche dwells in the internal world of humans and it has many components. For the purpose of this study, I prepared a diagram as a mental construct of selected components of the psyche.

FIGURE 7.1
SELECTED COMPONENTS OF THE PSYCHE

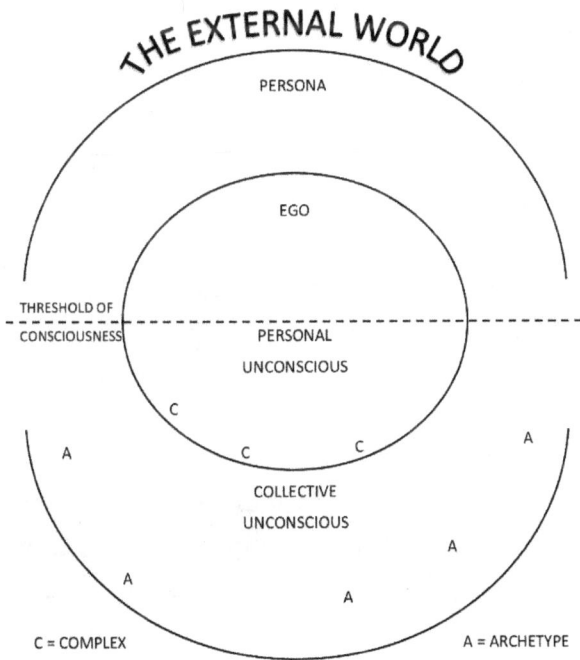

THE EXTERNAL WORLD

PERSONA

EGO

THRESHOLD OF
CONSCIOUSNESS

PERSONAL
UNCONSCIOUS

C

A

C

C

A

COLLECTIVE
UNCONSCIOUS

A

A

A

C = COMPLEX

A = ARCHETYPE

Figure 7.1 shows components of the psyche that I selected for this study:

- Conscious part of psyche
  - Ego
  - Persona
- Unconscious part of psyche
  - Personal unconscious
  - Collective unconscious.

The conscious realm is that part of the psyche of which we are aware and to which we have direct access. It is made up of the ego and persona. The ego is the center of consciousness. It manages our thinking about daily living. It directs our engagement of the external world. It also mediates interaction between conscious and unconscious realms of the psyche. It is the ego's responsibility to recognize psychological projections, withdraw them and integrate them into consciousness. The persona is the part of the psyche that functions like a mask or an avatar. It is how we present our self to society. The conscious realm of our psyche does not have direct access to the unconscious realm. Consciousness grows out of the unconscious realm of the psyche.

In the unconscious realm, there are the personal unconscious and the collective unconscious layers. The personal unconscious is a repository for content that was once conscious, but was repressed by the individual. The personal unconscious is also known as the shadow. Conscious content is repressed when it is unacceptable to consciousness. The personal unconscious contains complexes, which are emotionally charged entities. In "Complex, Archetype, Symbol in the Psychology of C. G. Jung" Jolande Jacobi explains that, because it is unconscious, a complex appears in a projected form as an attribute of someone or something in the external world.[3] Jacobi further explains that a complex is independent and acts like an autonomous personality; it has intentionality along with a drive, hopes, fears and desires.

The collective unconscious is a substrate that belongs to all humanity, while the rest of the psyche belongs to an individual. This substrate is a repository of the experience and potential of all humanity. The collective unconscious produces images, symbols, fantasies and dreams, all of which can be used to expand consciousness. The collective unconscious is populated by archetypes that are structural components of the psyche. In "Complex, Archetype, Symbol in the Psychology of C. G. Jung" Jolande Jacobi explains that archetypes are known through their projections to the outer world.[5] She explains that archetypes are like hidden organizers of life with connections to universal experiences that are deep in the psyche.

It is the ego's responsibility to mediate the interaction between the conscious and unconscious components of the psyche. In performing the role of mediator, the ego engages in activities such as these:

- Resolve the tensions between conscious and unconscious realms of the psyche.
- Process the emergence of content from the collective unconscious.
- Translate, interpret and amplify the emerging unconscious content for use by consciousness.
- Facilitate the evolutionary progress of the psyche by managing psychological projections.
- Recognize psychological projections, withdraw them and integrate them into conscious awareness.
- Expand conscious awareness.

Mediating the interaction between conscious and unconscious components of the psyche is a complicated role for the ego. The complication arises from the structure of the psyche. Analytical Psychology identifies four known functions of the psyche:[5]

- Thinking function
- Sensing function
- Intuiting function
- Feeling function.

Of these four functions, one is usually dominant, while the others are less developed or latent in the psyche. The dominant function is accessible to scrutiny by the ego because it is in consciousness. The other functions are not directly accessible by the ego because they are in the unconscious part of the psyche. The dominant function for an individual could be any of the four functions. The difficulty this creates for the ego is that it must adjust the mediation between consciousness and unconsciousness, when latent functions become known to consciousness.

FIGURE 7.2

POSITIONS OF FUNCTIONS IN THE PSYCHE

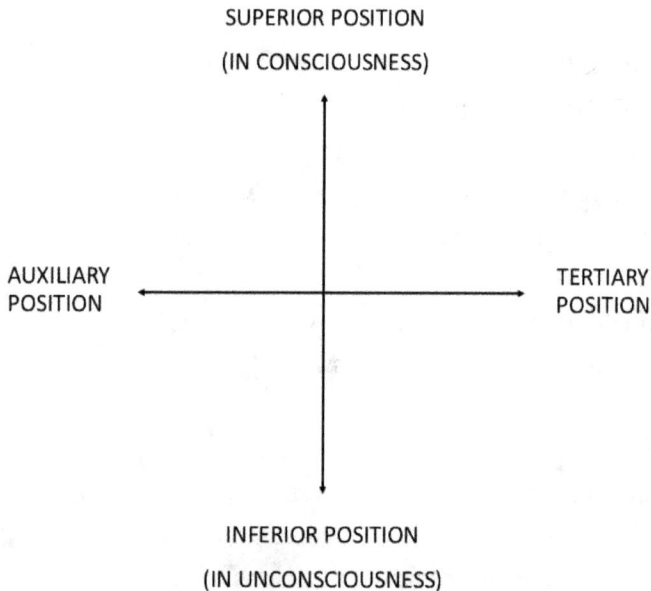

SUPERIOR POSITION

(IN CONSCIOUSNESS)

AUXILIARY
POSITION

TERTIARY
POSITION

INFERIOR POSITION

(IN UNCONSCIOUSNESS)

Figure 7.2 shows four positions occupied by the functions of the psyche. In "Energies and Patterns in Psychological Type", John Beebe provides definitions of the terms "superior", "auxiliary", "tertiary" and "inferior" as descriptors of the functions in the four positions in the psyche.[6]

- The function in the superior position is the most developed function and it is in the conscious part of the psyche.
- The function in the inferior position is the least developed function and it is in the unconscious part of the psyche.
- The function in the auxiliary position is like the right hand (of a right-handed person) and it is associated with competence.
- The function in the tertiary position is like the left hand (in a right-handed person) and is associated with vulnerability.

Beebe sees the axis formed by the superior and inferior functions as the 'spine' of the personality, and the axis formed by the auxiliary and tertiary functions as the 'arms' of the personality. An awareness of the functions that form the spine enables a person to form an identity and helps in determining how to relate to others. In Beebe's experience, the superior and auxiliary functions may develop naturally in childhood, but the tertiary and inferior functions do not appear until adulthood.[7]

The psyche evolves by interacting with the external world. The evolutionary mechanism for that interaction is psychological projection. A psychological projection can be either a defense against anxiety, or an emergence of content from the collective unconscious.

FIGURE 7.3

PROJECTION OF IMAGES TO THE EXTERNAL WORLD

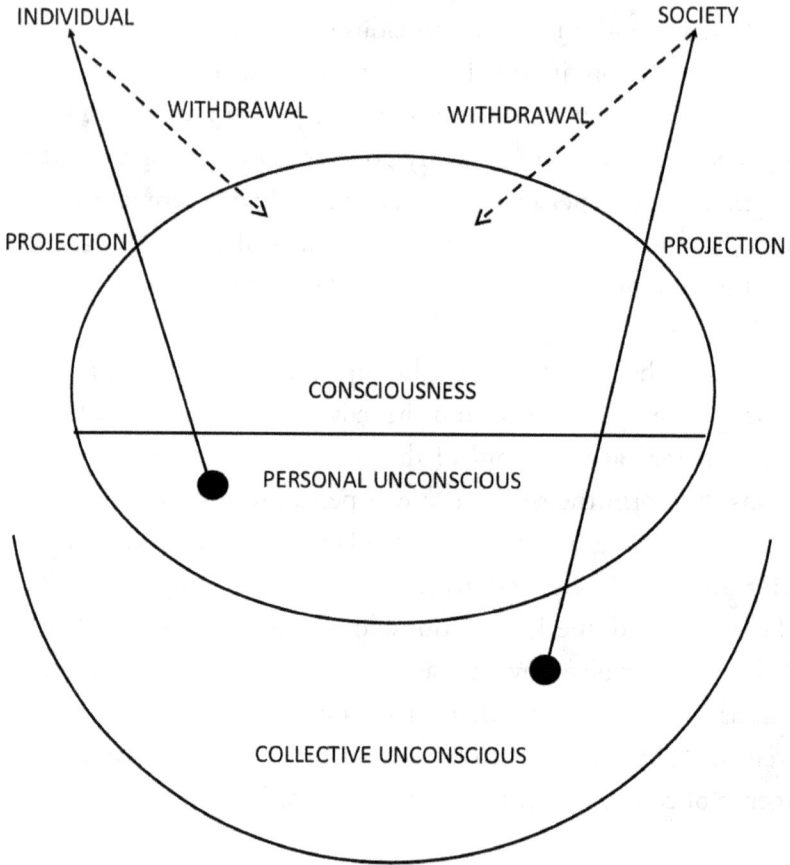

INDIVIDUAL

SOCIETY

WITHDRAWAL

WITHDRAWAL

PROJECTION

PROJECTION

CONSCIOUSNESS

PERSONAL UNCONSCIOUS

COLLECTIVE UNCONSCIOUS

Figure 7.3 shows examples of two projections. On the left is a projection from the personal unconscious to an individual in the external world. On the right is a projection from the collective unconscious to a society in the outer world. The arrows on dotted lines show the projections being withdrawn into consciousness. Psychological projection is the means by which content from the unconscious realm comes to the attention of the conscious realm of the psyche. It is an involuntary casting of an image from the unconscious realm of the psyche to a carrier in the external world. The carrier is an entity in the external world, for example, the universe, the earth, an institution, an object or a person. The carrier forms a hook for the projected image because it has something in common with the image being cast. The carrier may have a similarity to the image or some resonance with the content of the image. Projections are carried by entities in the external world. Something, or someone in the external world becomes a hook on which the projection hangs. According to Andrew Samuels, et al., in "A Critical Dictionary of Jungian Analysis" a psychological projection is a defense mechanism used to avoid the anxiety that is provoked when one is forced to face faults, weaknesses or destructive tendencies.[8] The projection can be normal, or it can be pathological. Overall, projections are a necessary part of normal psychological development and are the means by which we gain awareness of unconscious content.

The external world serves the psyche by providing the stimulus that activates a projection. When a social group, or an individual, moves too far from the foundations of reality, it triggers a response from the unconscious realm of the psyche. Carl Jung informs us in "The Archetypes and the Collective Unconscious" that collaboration between the conscious and unconscious realms is intelligent and purposive.[9] When the unconscious realm responds to a conscious attitude, it is compensatory in an intelligent way, as if trying to restore lost balance. In keeping with the functioning of organic systems, the psyche seeks to maintain balance. When there is a disequilibrium, the psyche has the means to achieve balance. Achieving balance involves significant effort and challenging decisions on the part of the ego.

In this study, I select psychological projections to demonstrate the evolution of the human psyche. The projections I choose are all initiated in the collective unconscious and they are all projected onto physical entities: the land, the earth, the universe, society and technology. These are projections from the collective unconscious of humanity as a whole and they occurred at milestones of human history that are characterized as revolutions due to the magnitude of their disruptions.

NOTES

1. See Carl Jung's definition of the psyche in the book "Psychological Types". Collected Works Volume 6, Paragraph 797.
2. See description of the psyche in "Structure & Dynamics of the Psyche" in Collected Works of C. G. Jung, Volume 8.
3. See description of a complex on pages 6 – 30 of "Complex, Archetype, Symbol in the Psychology of C. G. Jung" by Jolande Jacobi.
4. See description of an archetype on pages 31 - 73 of "Complex, Archetype, Symbol in the Psychology of C. G. Jung" by Jolande Jacobi.
5. See Carl Jung's functions of the psyche in the book "Psychological Types". Collected Works Volume 6.
6. See detailed descriptions of the psychological functions on pages 134 – 138 of "Energies and Patterns in Psychological Type" by John Beebe.
7. See sequence od development of the psychological functions on pages 34 of "Energies and Patterns in Psychological Type" by John Beebe.
8. See information about projections on pages 113 – 114 in "A Critical Dictionary of Jungian Analysis" by Andrew Samuels, Bani Shorter and Fred Plaut.
9. See information about the collaboration between consciousness and unconsciousness in "The Archetypes and the Collective Unconscious", Volume by Carl Jung. Collected Works, Volume 9, Paragraph 505.

Chapter 8

———— ∾∾ ————

Psyche's Response

In this chapter, I explain that Singularity's threat of human subservience to technology is unlikely to be realized because the human psyche does not stand still. The psyche evolves. My opinion is that technology evolves in tandem with the evolution of the human psyche, in a collective sense. I demonstrate this in periods in our history that are known as revolutions:

- The Agricultural Revolution
- The Scientific Revolution
- The Industrial Revolution, and
- The Digital Revolution.

At each revolution, I support my opinion that technology evolves in tandem with the psyche by demonstrating the following:

1. There is a change that occurs in the psyche during a revolution. This change involves relinquishing a collective human conviction that turns out to be an illusion and replacing it with knowledge founded in reality.
2. There is a corresponding change that occurs in technology during the revolution. This change involves offloading human knowledge to technology.

3. The relationship between the change in psyche and the change in technology shows that the human psyche and technology evolve in tandem.

There are instances where some individual humans do surrender their intelligence to technology, but humankind, as a whole, does not. We have become so accustomed to relying on the ATM's, smart phones, downloaded "apps" and the instrumentation on the dashboard of our cars, that we do not question their accuracy. All that means is we have offloaded certain rational tasks to technology. If technology is to outsmart humans, it will need to add non-rational capabilities to its repertoire. Examples are emotional intelligence, creativity and the functions of the psyche.

In summary, here are psyche's responses to Kurzweil:

- Yes, there are scenarios where rogue technology or hackers can wreak havoc with human affairs. That has always been so and will continue to be so. For technology to surpass human intelligence, technology must first discover what constitutes human intelligence. That necessitates the inclusion of psychology among the disciplines that currently support Artificial Intelligence. Cognitive Psychology is not enough because it pays attention to cognition, but does not cover the unconscious aspects of the psyche. Analytical Psychology is more inclusive; it covers both conscious and unconscious aspects of the psyche.
- The human psyche is continuously evolving. Historically, the human psyche engages the environment, learns from the engagement and acquires new knowledge. The acquisition of new knowledge produces a restructure of the psyche, enabling it to evolve to a higher level of capabilities. To surpass human intelligence, technology must take its evolution into account.
- When the human psyche reaches a knowledge overload, it offloads to software and technological devices in the

environment. That offloading frees up mental resources thus enabling the human psyche to perform at a more sophisticated level. The evolution of the human psyche proceeds by successive offloads to technology. For example, human capabilities were offloaded to the calculator, the computer, programming languages, algorithms, Artificial Intelligence (AI) and Machine Learning (ML) systems. The offloading increases in complexity over time. With each offloading of knowledge to technology, humans moved on to explore more complex capabilities.

- So far, the offloading of mental capabilities to technology has been mainly in the area of rationality ... logical thinking about that part of our knowledge which we are able to articulate explicitly. According to Analytical Psychology, there are other mental capabilities in addition to logical thinking; they include intuition, sensation and feeling.[1] We have not yet offloaded intuition, sensation, or feeling capabilities to technology. Such offloading would be necessary if, as Kurzweil predicts, technology is to surpass human intelligence.

- Different branches of Artificial Intelligence can extract knowledge from data produced by humans. For example, Machine Learning is a branch of Artificial Intelligence. Big Data refers to large repositories of data accumulated from situations such as online transactions or collection of medical records. Machine Learning is able to mine Big Data to discover the implicit knowledge that we possess, but which we are not necessarily able to articulate. Machine Learning uses Big Data to learn and iteratively improve its learning. However, mining Big Data does not extract motivation or intention that went into producing the Big Data. Both motivation and intention are elements of human mental capabilities that technology would have to master if it is to outsmart humans.

- There are historical revolutions which provide evidence that the human psyche evolves. They include the Agricultural, Scientific,

Industrial and Digital Revolutions. With each revolution, the human psyche acquires higher level capabilities. At each revolution, the psyche offloads knowledge to technology. The psyche evolves in tandem with technology. Singularity is not likely to enable technology to surpass human capabilities because technology and humans evolve together.

In essence, Singularity is about a prophesy that technology will match and exceed human intelligence by the middle of the twenty first century. The proponents of Singularity include emotion and consciousness under the umbrella of human intelligence, but they have not yet digitized those areas of human intelligence. We are now in the second decade of the twenty first century, and technology still has a lot of catching up to do in order to match human intelligence. Whole brain emulation is still experimental. The definition of consciousness has eluded the technological visionaries. Emotional intelligence is still beyond the grasp of those who believe that reverse engineering the brain is a precursor to creating a mind. If technology is to exceed human intelligence, it must first catch up with it. My observation is that while technology is advancing at high speed, it is not matching human intelligence. That is because human intelligence is also advancing. With each historical revolution, the psyche overcomes a hurdle of acquiring new knowledge and capabilities. The evolution of the human psyche can be charted in a series of revolutions over the course of history. In the following sections, I explain how the revolutions impact the evolution of our psyche.

8.1 Psyche Offloads Knowledge to Technology at each Revolution

Humans evolve through periods of order interrupted by chaos. Chaos takes the form of periodic disruptions that define our history. Historians characterize the mix of order and chaos as revolutions.

Examples of revolutions that have had significant impact on the human psyche are:

- The Agricultural Revolution
- The Scientific Revolution
- The Industrial Revolution, and
- The Digital Revolution.

These revolutions begin with a psychological projection that is characterized by a conviction which humans share about our place in the external world. When humans recognize that the conviction is an illusion, the psyche experiences a tension that disrupts its orderly functioning. The tension is due to the fact that the framework of conviction is recognized as being no longer serviceable, but there is no new framework to replace it. The tension in the psyche creates an opportunity for the emergence of content from the unconscious realm of the psyche to come into conscious awareness. By taking advantage of the new images and ideas that emerge, humans have the opportunity to acquire new knowledge and capabilities. The newly acquired knowledge enables us to establish a new framework for our relationship with the external world and stabilize our psyche. Over time, some of the new knowledge and skills that we acquire are offloaded to technology in the forms of tools, methods, software and machines. The offloading frees some of our mental resources and our time for other activities ... until the next revolution. With each revolution, the psyche evolves. One mechanism that makes that evolution possible is the psychological projection.

Philosopher Jean Gebser (1905 – 1973) wrote "The Ever-Present Origin" during the years 1949 through 1953. In that book, Gebser commented on the role of projection in the context of the creation of technology. He views the creation of technology as the externalization of the inner powers of the psyche to make machines, tools and laws.

The 1985 English translation by Noel Barstead and Algis Michunas states the following: [2]

> "Yet to the extent that the machine is an objectivation or an externalization of man's own capabilities, it is in psychological terms a projection. We have already spoken of the decisive role of projection in the emergence of consciousness; it is only because of these projections, which render externally visible the powers lying dormant within man, that he is able, or more precisely, that it is possible for him to become aware of this intrinsic potentiality which is capable of being comprehended and directed.
>
> All 'making,' whether in the form of spell casting or of the reasoned technical, construction of a machine, is an externalization of inner powers or conditions and as such their visible, outward form. Every tool, every instrument and machine, is only a practical application (that is, also a perspectival-directed use) of 'inherent' laws, laws of one's own body rediscovered externally. All basic physical and mechanical laws such as leverage, traction, bearing, adhesion, all constructions such as the labyrinth, the vault, etc., all such technical achievements or discoveries are pre-given in us. Every invention is primarily a rediscovery and an imitative construction of the organic and physiological pre-given 'symmetries' or laws in man's structure which can become conscious by being externally projected into a tool."

In Gebser's view, projection plays a decisive role in the emergence of consciousness. A projection makes externally visible the powers lying dormant within the psyche. A projection also makes it possible for humans to become aware of that potentiality. The creation of technology is the externalization of the inner powers of the psyche. Technology has a history of tools that mirror the evolution of the psyche. Some of the tools are the farming plow, the telescope, the computer, the Internet and Artificial Intelligence.

The introduction of new technology is not just about performing activities with mechanical devices. It is also about altering cognition about those activities. A new technology can create cognitive dissonance, or emotional mismatch, in humans because the new activities violate our frame of reference. We have frames of reference that are constructed from past experiences, present knowledge and expectations of the future. Where the new technology fits our framework, we embrace it, but if the framework and the technology are an uneasy fit, there can be a disruption of our frame of reference.

In "The Nature of Technology", W. Brian Arthur quotes psychologists who portray our frame of reference as not easily dismantled.[3] The way that humans see the world is ultimately linked to the manner in which we define our relation to the world. We have a vested interest in maintaining consistency because otherwise, our identity may be at risk. Arthur's account of the creation of technology has a sense of ancestry. He uses the expression "combinatorial evolution" to explain the evolution of technology. Each new technology is created partly from past technologies. For example, the creation of the Internet grew from an ancestral repository of past technologies including search engines, and networks. Communication by e-mail became possible because the Internet was already in the ancestral repository of technologies.

Arthur's notion of combinatorial evolution can be applied to the biology and psychology of humans. In biology, we see the concept of combinatorial evolution in the way humans inherit genes. Each human inherits genes from the ancestral repositories of genes from just two parents. The biological attributes of a human are built up from the ancestral repositories of the parent's genes. With a similar argument, psychologist Carl Jung uses the word "archetype" to describe human evolution that has a sense of ancestry. In Jung's view, the human psyche has access to a collective unconscious in which archetypes form a repository of past human experiences.[4]

Arthur proposed an ancestral repository of technologies that have potential use in future technologies. Jung proposes an ancestral repository of human experiences that informs future generations. Arthur's repository is a physical, tangible repository. Jung's repository is non-physical, unconscious and intangible. Arthur's ancestral repository of past technologies parallel Jung's ancestral repository of human experiences. I use Gebser's view of projection and Arthur's view of ancestry as a way to lead into the Jungian description of psychological projection. In this study, I explain that psychological projection is a mechanism by which the human psyche evolves. I also explain that the psyche and technology are evolving in tandem. I demonstrate their parallel evolutions in the contexts of the Agricultural, Scientific, Industrial and Digital Revolutions. In the next section, I explain the stages of psychological projection.

8.2   Stages of Psychological Projection

One mechanism of the evolution of the psyche is psychological projection. We evolve by the involuntary projection of unconscious content from our psyche unto the external world. Analytical Psychology informs us that there is an ancestral repository of human experience in the unconscious psyche. The content of the unconscious psyche comes to the attention of the ego in the conscious mind by means of projection. A projection originates in the unconscious realm of the psyche and bypasses conscious awareness. An image is involuntarily projected from the unconscious realm onto a "carrier" in the external world. A carrier is some entity with which the image finds resonance. With reflection, the ego comes to recognize the projected unconscious content as coming from within. Psychological growth comes from recognizing the projection, deliberately withdrawing and integrating it into the conscious realm of the psyche. The following description of the stages of a projection draws on the works of psychologists Carl Jung and Andrew Samuels. Along with co-authors Bani Shorter and Fred Plant, Andrew Samuels wrote "A Critical Dictionary of Jungian Analysis" where they describe two types of psychological projections.[5]

1. A defense against anxiety, where difficult emotions and parts of the personality that are unacceptable to consciousness become attributed to another person or institution or external object to provide relief and a sense of well-being.
2. A means of growth, where contents from the unconscious world are made available to the ego-consciousness. The encounter between the ego and the unconscious contents has the potential for psychological growth.

The external world of persons and things serves the internal world by providing the raw materials that activate the projection. Individuals and objects in the external world act as "carriers" of the projection.

The concept of psychological projection had been published by Sigmund Freud in 1894 when he defined it as a defense mechanism. The definition in terms of defense mechanism is limited to psychological projection as an unconscious process of concealing from oneself drives and feelings that threaten to lower our self-esteem and provoke anxiety. That publication occurred during the Industrial Revolution, but the meaning of psychological projection was not well known to the global public. It was much later that Marie-Louise Von Franz extended that definition to include psychological growth. In her book "Projection and Re-Collection in Jungian Psychology", she included psychological growth by explaining that projection occurs wherever we come to the limit of what humans know of the external world.[6] She further explained that wherever known reality ends, that is where we touch the unknown, and that is when we project an image onto the external world.

This study focuses on projection as a means of psychological growth, by offering the idea that historical revolutions have been milestones in the evolution of both the psyche and technology. In their dictionary, Samuels, et al., base their definition of a projection on their interpretation of Carl Jung's work. They identify six stages in the existence of a projection. I take the liberty of adding a seventh stage to elaborate the emergence of new knowledge into consciousness as the psyche restructures and stabilizes after the retraction of a projection.

- Stage 1: Conviction about the External Environment

  In the first stage of a projection, humans harbor a conviction that certain attributes we ascribe to the external world are in fact characteristics of the external world. We share the inner conviction because it fits with our frame of reference about the outer world in which we live.

- Stage 2: Recognition that Conviction is Flawed

  In the second stage of the projection, we gradually come to recognize that there is a difference between the external world and the attributes we ascribe to it. We recognize a discrepancy between our image of the external world and the actual external world. Awareness of this discrepancy creates a tension in our psyche. We no longer trust our conviction about the external world, but we do not yet have a new understanding of the external world.

- Stage 3: Assessment of Discrepancy

  In the third stage of the projection, the more reflective humans engage in an assessment of the discrepancy between our image of the external world and the actual external world. Our psyche finds it necessary to embark on the assessment because living with the discrepancy makes the tension of the psyche worse.

- Stage 4: Conclusion that Projection Was an Illusion

  In the fourth stage of the projection, we come to a conclusion that the attributes we ascribe to the external world do not actually belong to the external world. Our projected image does not match the external world. We see evidence that our conviction is an illusion. That evidence breaks the projection, but it creates turmoil within the psyche. Now, there is an urgent need for a new understanding of the external world. Our uncertainty about the external world feeds the tension in our psyche.

- Stage 5: Search for Origin of Projection

  In the fifth stage of the projection, reflective humans engage in a deliberate search for the origin of our projected image. In this search, we come to see the external world for what it is and realize that the projected image originates within us. The realization that we harbored an illusion increases the tension in our psyche. We experience the pull of the tension between our old frame of reference, the projection to which we had become habituated, and the new frame of reference, which is not yet formed.

- Stage 6: Retraction of Projection

  In the sixth stage of the projection, those humans who recognize the origin of the projection are in a position to retract it. We retract the projected image from the external world by taking responsibility for the projection having originated within us. The effect of the retraction is that the tension in the psyche increases. We experience a tension between two frames of references. At one end of the tension, we are relinquishing our old frame of reference, which we now know to be inaccurate. At the other end of the tension, we are uncertain of what the new frame of reference will be. We have to continue living in the external world, but we experience tension because we are not sure how to relate to it. Our ego is at a loss about how to conduct our personal lives in the context of the external world.

- Stage 7: Emergence in Restructured Psyche

In the seventh stage of the projection, the tension within the psyche creates an opportunity for the emergence of content from the unconscious aspect of the psyche. New images and ideas emerge from the unconscious psyche. If we are able to translate, interpret or amplify the emerging content, it helps us create a new frame of reference for interacting with the external world. If we are able to integrate the emerging content into consciousness, it brings about a restructuring of the psyche. Our new frame of reference of the external world coupled with our restructured psyche, enable us to explore possibilities for new methods and tools. Those methods and tools that work well in our new frame of reference find resonance with the psyche and they endure. Gradually, we populate the new frame of reference with methods and tools derived from exploration of the content that emerged from the unconscious psyche. Over time, humankind integrates the emerging content into a coherent new frame of reference of the external world. The integration of emerging content brings about a restructuring of the psyche. The result is an expansion of consciousness that occurs in support of the evolution of the psyche. As consciousness expands with new knowledge, we get to the point where our psyche experiences an overload of knowledge and skills. We offload our knowledge to technology in the forms of machines, methods, laws and tools. By means of psychological projection, technology evolves in tandem with our psyche.

8.3   Psyche & Technology in the Context of Evolution

One of the characteristics shared by the human psyche and technology is complexity. They are both structurally complex and behaviorally complex. One way to examine both in the context of evolution is to use Chaos Theory. Chaos Theory describes the behavior of complex systems. Psyche and technology both qualify as complex systems because they are made up of many interacting variables along with several feedback loops. According to Chaos Theory, complex systems sometimes behave in an orderly manner, and other times in an unpredictable manner.[7] Past behavior does not bear a causal relationship to future behavior. As complex systems, the psyche and technology have decentralized control. Each has multiple components that combine to produce resulting behavior. Their structure is decentralized and that makes them resilient and adaptive. Each component in the complex system has the latitude to quickly respond and adapt independently to changing events. Complex systems have a surprise-generating mechanism known as emergence.[8] Emergence is a distinguishing feature of complex systems. Emergence is a surprise output, or novelty, generated by a complex system when the interactions among multiple variables reach a high level of complexity. In describing the evolution of the psyche and technology, I apply the concept of emergence.

In the next four sections, I apply the stages of a psychological projection to historical revolutions that are milestones in the evolution of both the psyche and technology:

- Agricultural Revolution
- Scientific Revolution
- Industrial Revolution
- Digital Revolution.

8.4   Agricultural Revolution: Persephone, Demeter & Hades

In this section, I select the story about the discovery of the seasons of the year to demonstrate the evolution of the psyche through psychological projection. The early Agricultural Revolution provides one example of how technology evolves in tandem with the psyche. From the eighth century BC through the second century AD, several countries transitioned from their nomadic hunting and gathering practices to farming and settlement in specific locations. At that time in history, literacy was not widespread. Few humans knew how to read and write. Although early forms of calendars were being developed, they were not readily available to nomadic tribes. Humans were accustomed to being sustained by hunting animals and gathering edible foods from trees. We held the conviction that the hunter-gatherer lifestyle would continue indefinitely. Food would always be available to us; all we had to do was find it. We felt it would always be so because reliance on the earth's bounty was the only life we had known. That conviction prevailed for years. Then, as the nomadic populations grew in size, the available food began to diminish. As food sources diminished, it shook our conviction that the nomadic lifestyle was sustainable.

Slowly, we realized that there was a growing difference between our conviction and the availability of food in the external world. Where there used to be a lot of food on trees and many animals to hunt, we came to recognize that our growing population was depleting the earth's bounty. We became aware of a growing discrepancy between the available food and the amount of food necessary to feed our population. There was a mismatch. Our frame of reference no longer fit. Having to live with that discrepancy created a heightened interaction among variables in our psyche. Hunger could no longer be satisfied by hunting. Nomadic travelling was not enough to feed the tribes. The discrepancy also activated more feedback loops between our psyche and the environment. Feedback loops from the environment informed us of diminishing sources of animals to hunt and fewer trees bearing food. We no longer trusted our conviction about the external world being an unlimited supplier of food, but we did not have any other way of obtaining food. When the animals and the edible foods grew scarce, survival of nomadic tribes was threatened. Due to the heightened interaction of components and feedback loops of our psyche, we began to notice that there was a cycle of activities in nature. There were periods of time when food was plentiful and other periods when food was scarce. However, there was no shared understanding of what caused food to be plentiful or scarce.

In an attempt to explain how food becomes available in nature, we composed a myth. By "myth" I do not mean falsehood; I mean an allegorical story created to facilitate a shared understanding in a society of pre-literate, nomadic tribes. In the early years of the Agricultural Revolution, a myth was a story composed to enable pre-literate humans achieve some common understanding of changes that humans observed in nature. We imagined there to be gods and goddesses who control nature. The allegorical story that I am about to relate arose from our imagination as we tried to understand changes that we noticed in the environment. Unable to read or write, we composed the story to facilitate a shared understanding of the seasons in nature. This story involves gods and goddesses to whom we ascribed roles associated with the events that we noticed about times when crops were scarce and when they were plentiful.

The story of Demeter, Persephone and Hades relates the mythological creation of the seasons of the year.[9] Our human imagination created Demeter as a goddess of fertility, grain and agriculture. We gave her the role of making things grow and flourish. In the story of Demeter, we identified her daughter named Persephone.

One day, Persephone was out in a field of flowers enjoying the light and warmth of the day, while picking flowers. Suddenly, the earth opened up beneath and out of the opening, rode Hades the god of the Underworld on a chariot. Hades had fallen in love with Persephone. Knowing that Demeter would not approve his marriage to her daughter, Hades abducted Persephone and rode off with her to the Underworld where he lived. As Hades rode into the Underground, the opening in the earth closed. Demeter searched for Persephone but could not find her. Distraught about her missing daughter, she neglected her role as goddess of fertility and devoted her days to locating her missing daughter. The grains and other plants soon withered. As food came to be in short supply, an increasing number of hungry people appealed to the gods to help Demeter find Persephone. The gods helped Demeter to locate Persephone in the Underworld with her abductor, Hades. Demeter begged Hades to let Persephone return to the surface of the earth, but he refused, claiming that she was now his wife and queen of the Underworld. Knowing that if she ate anything in the Underworld, she would not be able to leave the Underworld, Persephone had been careful not to eat anything. However, Hades tricked her into eating seeds of a pomegranate. A dispute arose between Demeter and Hades about whether Persephone should leave the Underworld and go back to the surface of the earth. To resolve the dispute, other gods were asked to intervene. The gods decided that each year, Persephone should spend a portion of her time in the Underworld with Hades, and a portion of her time with Demeter on the surface of the earth. The story is an allegory about planting seeds and growing crops as the seasons change throughout the year. Since this story was shared by word-of-mouth across many tribes, there are many versions. Here, I offer a general interpretation of the story as it came to depict seasons of the year.

- SUMMER was the time when Persephone picked flowers in the light and warmth of the field on the surface of the earth.

- AUTUMN was the period of time when crops and plants faded away, and sunshine gave way to darkness while Demeter searched for Persephone.
- WINTER was the time that Persephone spent in the Underworld, while trying to avoid food.
- SPRING was the period when Persephone returned to the surface of the earth with Demeter, when the sun shone again, crops sprouted and plants flourished.

Pomegranate seeds were symbols of fertility; they represented the seeds that are planted in the ground with the expectation that they will sprout in spring. That is the story of how the nomadic-turned-farming tribes identified seasons of the year as a context for the growth of crops. The story was shared by word-of-mouth, across tribes as they settled into sedentary life to establish agricultural communities.

Those humans who saw the usefulness of the story of Persephone may have moved on to embrace farming life enthusiastically, without looking back to compare the old model of nomadic life with the new model of farming life. Those who did look back were the ones who wanted to find out why there was a difference between our conviction about the earth as an unending source of food and the realization that food in the external world could actually diminish. Those who looked back came to the conclusion that the earth does not guarantee an unlimited supply of food. Our projected image of having an unlimited food supply did not match the circumstances in the external world. In the earth's depleting food supply, we saw evidence that our conviction about an everlasting food supply was an illusion. That evidence broke our projection. The more reflective humans began a search for the origin of our projection. How did we come to assume that the earth's bounty would be an everlasting source of ready-made food for us? Why did we not notice earlier that when seeds fell to the ground, they grew into crops? In this search, we came to understand that the external world as a food source that can be depleted, if it is not replenished periodically. We also realized that the projected image of an unlimited food supply originated within us.

Regardless of the degree of interest in discerning the origin of our projection or retracting our projection, there was enough interest in moving on from the old model of nomadic life to farming life. The story of Demeter created an opportunity for emergence of content from the unconscious realm of the psyche. New ideas began to emerge from the unconscious psyche. We took advantage of the new ideas to explore new ways of obtaining food from the external world. We experimented with planting seeds. We learned how to grow plants. Over the course of the Agricultural Revolution, we selected crops that could yield high enough volume to keep our communities alive. We learned how to pay sustained attention to the calendar and marked out seasons of the year. We worked out how to discipline ourselves into the rhythm of labor necessary to grow grains according to the seasons of the calendar. We paid attention to the lifecycles of animals. We determined which animals could be domesticated. We selected animals that could reproduce at a speed which would be adequate to supply us with meat for our survival. Some humans worked out methods for farming; others created tools to enhance farming. Those methods and tools that worked well on the stationary farms were retained in the establishment of a new frame of reference about the external world.

Over time, humankind discovered which new ideas were useful and integrated the emerging content into a coherent new frame of reference for relating to the external world. The increased interaction among the components of our psyche and the feedback loops between psyche and external world produced the emergence of the imagination to compose a story that enabled the definition of seasons of the year. The ideas and images that emerged from the unconscious helped to us to generate new ways of relating to the external world. We realized that we had to take control of creating our food sources, storing and managing our supplies. As our agricultural skills increased, we became more confident that we were in charge of our agricultural future, and our psyche achieved a new level of functioning. This new level was the result of a restructuring of our psyche. The restructuring came about in an expansion of consciousness, based on our creative use of images and ideas that emerged from the unconscious, while we used the Demeter story as a point for launching a new frame of reference. One outcome of the restructuring was the increased agricultural capabilities in the conscious part of our psyche. Another outcome was that many of the new agricultural skills were subsequently offload to technology. We created calendars defined by seasons of the year. We scheduled times for planting and harvesting various crops. We learned the reproductive cycles of animals grown for food, and we selected animals for domestication. Using the allegorical story, we created about Demeter, we were able to structure a new frame of reference about the earth as a source of sustenance. With that story, enough humans learned to take charge of our agricultural future by using knowledge of the seasons for growing crops.

The advocates of Singularity may view the Agricultural Revolution as an instance of technology outperforming humans in agricultural tasks. They would be right. Agricultural tools outperformed humans in speed and reliability. However, I should point out that technology did not evolve any further than humans. Humans created new agricultural knowledge, then built that knowledge into agricultural tools and methods. By offloading agricultural capabilities to technology, humans enabled the evolution of technology. The offloading also allowed humans to divert resources from labor-intensive agricultural tasks and focus on tackling higher level capabilities. The Agricultural Revolution was one step in the evolution of Homo Sapiens. During that revolution, technology did not surpass, or even match the intelligence of humans. Technology evolved in tandem with the human psyche.

8.5   Scientific Revolution: Copernicus, Rheticus & Galilei

In this section, I select the creation of the heliocentric model of the universe to demonstrate the evolution of the psyche through psychological projection. The Scientific Revolution offers another instance of how technology evolves in tandem with the psyche. When this revolution started, there was no technology related to heliocentricity or the laws of gravity. The Scientific Revolution consisted of the creation and socialization of empirical methods of engaging the external world. There were changes involved in letting go of the earth-centered view of the universe, while creating a new sun-centered frame of reference. Between the sixteenth and mid-nineteenth centuries, humans paid special attention to the empirical nature of the universe. There were conjectures about the planets in the universe. There were calculations to support, or disprove, the conjectures. Astronomers and mathematicians were composing hypotheses about the orbits of planets. Their hypotheses were based on their intuition and what their unaided eyes could see of planetary orbits.

Before the Scientific Revolution, religion held sway. Humanity held the conviction that we were privileged creatures, situated in the center of God's universe. We assumed that was our rightful place since we believed Homo Sapiens to be God's proudest creation. That was enough to tell us we are located at the center of the universe. If we needed further affirmation, the rising and setting of the sun were considered physical confirmation that we were living on a stationary earth, orbited by the sun. The prevailing view was a geocentric universe that had been constructed by Claudius Ptolemy, who lived in the second century. In a collective psychological projection, we humans had cast an image of geocentricity onto the universe.

I am going to relate a legend involving three people who played significant roles in enabling humans to recognize that our conviction about a geocentric system was flawed. They are Nicolaus Copernicus, Georg Rheticus and Galileo Galilei. Copernicus lived in the period 1473 to 1543. He was born in Poland, where he lost his father while he was a young boy. Copernicus' maternal uncle Lucas Watzenrode took charge of the boy's education and arranged for Copernicus to study at universities in Poland and Italy. Copernicus studied mathematics, astronomy, Catholic canon law and medicine. Copernicus later worked as his uncle's secretary and physician in a castle in Frombork, where Watzenrode was a Catholic Bishop. In his spare time, Copernicus pursued his interest in astronomy, making observations of the orbits of Mercury, Mars, Venus, Saturn and Jupiter. By the year 1514, he had drafted his proposal of a heliocentric system in "Commentariolus" ("Little Commentary"), which he shared with friends, but did not publish. By 1532, he had written "De Revolutionibus Orbium Coelestium" ("On the Revolutions of Heavenly Spheres").[10] His friends urged him to publish. The explanation he offered for not publishing was that he did not want to have to deal with the public reaction to such a strange notion as a heliocentric system. I have to wonder if he was not averse to publication because he anticipated a negative reaction from his employer, the Catholic Church. In 1539, Georg Rheticus came to Frombork to study with astronomers living there.[11] He was a mathematics lecturer on leave from the University of Wittenberg. He spent two years as Copernicus' student. At Rheticus' persuasion, Copernicus agreed to publish "On the Revolutions of Heavenly Spheres" and asked Rheticus to supervise the printing. While supervising the printing, Rheticus found that he had to leave Frombork to return to his teaching post at the University of Wittenberg. He left the supervision of the printing to a Lutheran theologian, who could not bring himself to print a book about a heliocentric system at conflict with his own beliefs. Unilaterally, he decided to insert a page at the front of the manuscript. That page stated that the earth did not really orbit the sun; it was just an imaginative device to simplify the mathematics.[12] The publication occurred on the same day that Copernicus died in 1543. At this point in time some humans recognized that our projection of a geocentric system was flawed. They were mostly astronomers and mathematicians. The bulk of humanity would recognize the flaw in later years.

Galileo Galilei is another important figure in this legend. He lived in the period 1564 -1642. Galilei studied at the University of Pisa and later became a lecturer in mathematics. He lived in Italy where he practiced astronomy and physics. Having studied Copernicus' publication of "On the Revolutions of Heavenly Spheres" about a heliocentric system, Galilei found it compatible with his own astronomical observations. His endorsement of Copernicus' heliocentric system put him at odds with the Catholic Church. In 1615, the Roman Inquisition had reviewed the publications about the heliocentric system, found them to contradict scripture, and therefore decided that it was heretical for anyone to promote such a system. When confronted by those who regarded the heliocentric system as heresy, Galilei made the distinction that the heliocentric system is about science, while scripture is about faith and morals. In the year 1616, the Inquisition had determined that heliocentricity was heresy and Copernicus' book was placed on the list of prohibited reading. Pope Paul V ordered Galilei to abandon his endorsement of heliocentricity as truth. In 1632, Galilei published a book titled "Dialogue Concerning the Two Chief World Systems" which came to the attention of the Inquisition.[13] After being interrogated by the Inquisition, Galilei was sentenced to house arrest for the remainder of his life and his book was banned. While under house arrest, he decided that experiments were necessary to demonstrate evidence of a heliocentric system. He experimented with the construction of a telescope, or spyglass, as it was known then. When he produced a serviceable telescope, he demonstrated it to people who ordered more telescopes. Shippers bought telescopes to observe approaching ships on the horizon. Astronomers bought telescopes to observe the movement of planets in the sky. Merchants bought telescopes for trading purposes. Construction of telescopes created a source of income for Galilei, while he was under house arrest. In his way of promoting heliocentricity, Galilei built telescopes, which provided evidence that the earth moves around the sun. That evidence eventually broke our projection.

Over a long period of time, humans gradually came to recognize that the geocentric system was an illusion. Astronomers and mathematicians led the way. Others were slow to follow. The Catholic Church placed Copernicus' "On the Revolutions of Heavenly Spheres" on the list of prohibited reading from 1621 to 1835, a period of over two hundred years. Advocates of a heliocentric system were deprived of publications that explained the topic, discouraged by the Inquisition, placed under house arrest and excommunicated for heresy. Those actions provided a strong disincentive to pursue any curiosity about heliocentricity.

The notion of a psychological projection originated with the birth of western psychology in the nineteenth century. The idea of searching for the origin of a projection may have begun at that time. However, the evolutionary growth of humans did not start with psychologists. Long before psychology became a profession, philosophers were noticing that humans create mental models of the external world. The less philosophic among us would likely have embraced the heliocentric model without looking back to wonder about the origin of the geocentric model. The more philosophic among us would likely have wondered about the origin of the geocentric model. While conducting a mental search for the origin of our projection, we would have raised questions. If God is omnipotent, why did he allow us to live under an illusion? Was God under the illusion as well? Could we continue to believe in God when the astronomers seem to be able to discover more from their empirical methods? Eventually, some realized that the geocentric universe is our projection that originated within us.

Regardless of how many of us retracted the projected image of geocentricity from the universe, there were enough humans embracing the heliocentric model for it to become our new frame of reference. The new frame of reference was derived from exploration of the content that emerged from the unconscious psyche. We realized that we had to take control of learning about our universe and determining how to function in it. As we learned more about the heliocentric universe, we became more confident that we were in charge of our scientific future.

A restructuring or our psyche occurred based on our creative use of images and ideas that emerged from the unconscious part of our psyche. One outcome of the restructuring was that humans acquired new scientific capabilities. When we mastered the new scientific methods, we offloaded them to technology. As the Scientific Revolution progressed, the human psyche offloaded scientific knowledge to technology in the form of scientific methods and skills such as crafting lenses, building telescopes and assembling cameras. The Scientific Revolution was one in a series of achievements in the evolution of Homo Sapiens. By offloading the new knowledge to technology, we freed resources of our psyche to tackle the challenges of the next revolution. During the Scientific Revolution, the interplay of the feedback loops from external world to psyche generated the emergence of the empirical methods and tools that established the heliocentric model as superior to the geocentric model. The images and ideas that emerged from the unconscious part of the psyche helped to us to generate new scientific ways of relating to the external world.

Those who promote Singularity will notice that, during the Scientific Revolution, technology outperformed humans in activities related to science. Technology performed empirical tasks faster and more reliably than humans. However, technology did not evolve any further than humans did. Technology advanced, but did not match human intelligence. The technologies of science were created by humans offloading scientific knowledge to tools. By offloading scientific knowledge to tools, humans enabled technology to take over some of the manual and mental tasks. Technology and the psyche evolved in tandem. Humans built scientific knowledge into tools to enable technology to take over human responsibilities. That freed us to give attention to more demanding activities. By offloading scientific knowledge to technology, humans freed up mental and physical resources for more sophisticated activities, and the next revolution.

8.6   Industrial Revolution: Cugnot, Benz & Ford

In this section, I select the manufacture of automobiles to demonstrate the evolution of the human psyche through psychological projection. I also show how technology evolved in tandem with the psyche. During the Industrial Revolution, there were several inventors from many countries working independently on the creation of an automobile. To illustrate the joint evolution of the human psyche and technology, I choose the work of three inventors: the Frenchman Nicolas-Joseph Cugnot, the German Karl Benz and the American Henry Ford. Historians generally agree that in 1769 Cugnot created the first steam-powered automobile that was capable of transporting humans.[14] In 1885, Benz produced the first gasoline-powered automobile.[15] Ford created the first mass-produced automobile in 1908, then followed up in 1913 with the establishment of a moving assembly line for producing automobiles.[16] I use the work of these three inventors to demonstrate the evolution of the psyche and the accompanying evolution of technology.

The Industrial Revolution gives us an example of how technology evolved in step with the psyche from the eighteenth through the nineteenth centuries. In the early years of the revolution, we regarded energy and the power necessary to do work as belonging to animate creatures, primarily humans and to a lesser extent horses. We had lived our lives in the understanding that we are responsible for doing work, with help from animals. Before the invention of the steam engine, transportation was accomplished by living creatures. People pulled carts. Animals hauled wagons. Humans rode horses. We did not anticipate that machines could be made to do work powered by manufactured energy. We did not expect that inanimate machines could take over the work being done by thinking humans.

When this revolution began, there was no technology about the types of energy that could power mass production or transportation. There were inventors who were experimenting with novelties that had the potential to become a steam engine, an electrical battery and a conveyer belt. Their experiments were based on their intuition about how to manufacture energy; how to store energy; and how to use the stored energy to power inanimate objects. When humans realized that our concept of energy as the power of humans to do work was being challenged, we experienced a great change in our psyche. There was accumulating evidence that energy did not belong exclusively to animate creatures like humans and horses. Experiments of inventors made it clear that energy could be commoditized. Energy could be manufactured, stored and later used to make inanimate machines do work. Steam could be boiled to create a steam engine. That was difficult for us to comprehend because our past experience told us that animate creatures possess energy to do work while inanimate creatures do not. The creation of energy to make machines do work was outside of our experience. Frenchman Nicolas-Joseph Cugnot, who lived in the period 1725 through 1804, built a steam-powered automobile.[17] Having served in the French Army and noticed the need for transportation, he chose to design a vehicle to transport guns and officers to the battlefield. Although the functionality of this self-propelled three-wheeled vehicle was demonstrated in 1801, the vehicle was not manufactured because King Louis XV showed no interest in funding the manufacture of automobiles. Due to the diversion of monetary resources during the French Revolution, Cugnot lost his pension. He discontinued his work on the steam-powered vehicle and lived in poverty in Belgium. Unfortunately, his poverty prevented him from pursuing better performing three-wheeled vehicles. Years later, Cugnot accepted Napoleon Bonaparte's invitation to return to France, but he died in 1804 before having time to manufacture automobiles. Cugnot's work made humans aware that energy did not belong just to animate creatures. Our conviction about energy was flawed. Energy could be created and stored. By building a steam-powered tricycle, Cugnot had demonstrated the use of energy derived from steam could be used to power a machine.

Karl Benz was a German engineer who was born in 1844 and died in 1929. In 1885, Benz produced a gasoline-powered, three-wheeled Motorwagen.[18] It had two seats, one for the driver, the other for a passenger. Benz's enthusiasm for improving versions of the Motorwagen flagged when he found that the local people regarded it more as a curiosity than as a functional vehicle for transportation. His wife Bertha decided to demonstrate the practical use of the Motorwagen by driving their two sons from their home in Mannheim to visit their grandmother in Pforzheim.[19] The distance between the two cities is over 60 kilometers. Bertha did not tell her husband until she had completed the trip, when she sent him a telegram. That trip made Germans aware of the practicality and safety of the Motorwagen. With his enthusiasm revived, Benz improved the features of the Motorwagen and began marketing it to the public. Benz's work demonstrated that energy does not belong exclusively to animate creatures. The human conviction that energy belongs to animate creatures was flawed. By building a gasoline-powered Motorwagen, Benz demonstrated that energy in the form of gasoline could be used to power an inanimate automobile.

Henry Ford (1863 – 1947) was an American automobile manufacturer.[20] Ford was not the inventor of the automobile, electricity or the assembly line. His innovation was to combine all three into a four-wheel vehicle for the mass transportation of people. At the time, cars were toys for the wealthy. Ford set himself a goal of building affordable cars for the masses. He built the gasoline-powered Model T car in 1908. Ford set out to build a car that was durable, easy to repair, low in cost and versatile enough to travel on rugged road surfaces. His efforts were rewarded; fifteen million of the Model T cars were sold. Ford then turned his attention to speeding up car production by establishing an assembly line. The moving assembly line brought the chassis along the conveyer belt so that workers could attach parts along the way. In 1913, Ford had completed the assembly line. He had added new features. For example, cars no longer needed hand cranks to start the vehicle because they were fitted with battery-powered starters. In addition, the assembly line reduced the time to produce a car from 12 hours to 2.5 hours. By building cars on an assembly line, Ford showed that energy does not belong exclusively to animate creatures. That was more evidence that our human conviction about energy belonging to animate creatures was flawed. Steam, gasoline and electricity could be stored as energy and then used to power inanimate machines.

The work of inventors Cugnot, Benz and Ford had all shown that energy was not an attribute that belonged exclusively to animate creatures. They had demonstrated that energy could be commoditized. To build automobiles, they had created energy and stored energy. There were steam engines, gasoline engines and electrical batteries. The stored energy was then used to produce self-propelled automobiles. Humans could see that our conviction about energy belonging to animate creatures was flawed. How could we have been so wrong? We pride ourselves on being the most intelligent species, Homo Sapiens. If consciousness is supposed to mirror the world, why did it take us so long? For centuries, we harbored a conviction that was at odds with the external world. Some humans might have taken the time to assess the discrepancy between our conviction and the reality in the external world. Perhaps the philosophers thought deeply about the discrepancy. Others probably hurried out to acquire new automobiles and go driving for the pleasure of it.

Following the inventions of Cugnot, Benz and Ford, humans had the opportunity to observe that vehicles could be made to provide transportation when powered by stored energy such as a steam engine, an electrical battery, a gasoline fed engine. We now had to contend with the reality that machines can be made to do work. We also had to face the real possibility that machines could replace us. Inventors had broken our projection by shifting the sources of energy from humans and animals to manufactured sources. Supplied with manufactured energy, machines could be made to do work. It was clear that our conviction about energy belonging to animate creatures was an illusion. Over the course of the Industrial Revolution, humankind generated new ideas about energy. We took advantage of new ideas that emerged from the unconscious psyche. While we did not all withdraw our projection from the external world, we integrated the new forms of transportation into a coherent new frame of reference for relating to the external world. At the time of the Industrial Revolution, the discipline of psychology was in its infancy. The concept of psychological projection had been published by Sigmund Freud in 1894 as a defense mechanism.[21] That publication occurred during the Industrial Revolution, but the meaning of psychological projection was not well known to the global public. It was much later that Marie-Louise Von Franz extended the definition of psychological projection to include psychological growth. Since psychological projection was not well known during the Industrial Revolution, it is unlikely that we humans would have used it to assess our flawed assumption about energy belonging to animate creatures. However, the evolution of the human psyche continued even though we did not yet have the psychological words to express our psychological growth. Those humans who understood the significance of projection in the context of psychological growth may have applied their knowledge in retracting the projection. Others may have simply embraced the new inventions and put them to use. Some may have focused conscious awareness on relinquishing an outdated world view in exchange for a more up-to-date world view. For the rest of humanity, we continued to marvel at industrial innovations and take advantage of the new modes of transportation, without necessarily comprehending the psychological implications of exchanging one framework for another. That approach enabled us to adapt to changes in the external world. Although the retraction of a psychological projection was not uppermost in our minds, the evolution of our collective psyche continued.

During the Industrial Revolution, the human psyche evolved. Consciousness was restructured to include new knowledge. We recognized that energy could be obtained from multiple sources in nature, for example, steam, gasoline and electricity. We came to accept that energy could be commoditized. It could be created and stored for later use, for example to power motor cars. We realized that machines could do work that had previously been done by humans. Not only that, the machines could perform tasks faster and with more fidelity than humans could. We learned not to take our assumptions for granted, but to be more observant about the alignment between our inner world assumptions and our observations about the external world. Then we brought the new technology about automobiles into alignment with the knowledge we acquired in the evolution of humans during the Industrial Revolution.

For the promoters of Singularity, the Industrial Revolution was another occasion when technology outperformed humans in activities related to industry. That is true. However, technology did not evolve any further than humans did. The technologies of industry were created by humans offloading industrial knowledge to tools. Technology and the human psyche evolved in tandem. Humans built industrial knowledge into tools to enable technology to take over human responsibilities. By offloading industrial knowledge to technology, humans freed up mental and physical resources for more sophisticated mental activities, and the next revolution. During the Industrial Revolution, technology was the recipient of knowledge that humans offloaded as methods and tools in the areas of transportation. Technology advanced, but did not match human intelligence.

The overall outcome of the Industrial Revolution was an emergence from the psyche of industrial knowledge and practices that enabled humans to develop automobiles and take charge of our transportation. We took advantage of the emergence of new ideas about locomotion to enable establishment of an industrial society. Technology evolved in tandem with the psyche in the sense that much of the industrial knowledge was offloaded to tools and methods that support industrial societies. The self-propelled automobile held a place of pride in the Industrial Revolution.

8.7   Digital Revolution: Shannon, Jobs & Jordon

For this revolution, I select three topics to demonstrate the evolution of the psyche through psychological projection. The selected topics are the digital computer, the personal computer and Machine Learning. I also show how technology evolved in tandem with the human psyche. To demonstrate the parallel evolution of the psyche and technology, I choose the work of three inventors: Claude Shannon, Steve Jobs and Michael Irwin Jordon.

At the dawn of the Digital Revolution, humans held the conviction that learning was the purview of humans. When this revolution began, there was no digital computers, and no technology about machines being able to learn on their own, without help from humans. Beginning in the late 1950s, we shifted our focus from mechanical and analogue technology to digital technology. We conducted our lives in the belief that humans are responsible for performing tasks that required mental skills. Of course, there were labor-saving tools such as the calculator, but there were no machines that could execute instructions written by humans. Before the invention of computers, mental tasks were performed by humans. We manually recorded data to keep track of grocery lists and bills. We calculated taxes. We maintained records of marriages and births and deaths. We did not expect machines to be able to perform mental tasks. We did not expect that inanimate machines could take over the work being done by thinking humans.

At the start of the Digital Revolution, the public had no technology about recording data in digits and then performing operations such as storing, sorting, calculating and producing reports. However, there were inventors thinking about building machines to perform just those tasks. There were inventors who were experimenting with novelties that had the potential to become thinking machines. They were trying to build computers that would perform mental tasks if humans produced logically sequenced, executable instructions written in code for machines to perform logical tasks.

Claude Shannon was an American computer scientist who lived from 1916 to 2001. He developed a mathematical communication model that would lay the foundations for the digital computers.[22] His master's thesis "A Symbolic Analysis of Relay and Switching Circuits" (1940) was regarded as one of the most significant theses of the twentieth century. In 1948, he built on the work of previous researchers to produce "A Mathematical Theory of Communication" in which he separated communication signals from the content of the message being transmitted. He showed that computers could be built using electronic relays and switches.[23] The basic design of the digital computer is founded on the notation that data can be represented in binary terms. "True" and "False" represent open and closed switches, while "0" and "1" represent digitized data. Digital coding is based on using binary digits, or bits, to represent information in two values: 0 and 1. The simplicity of using a system of two values enables fast processing of information. Shannon's theory rendered analogue information into binary digits that could be quantified, stored and transmitted between humans and machines. His theory laid the foundation for processing and transmitting information by digital computers. The early commercial digital computers were mainframes that business organizations used to process business transaction in financial systems such as payroll systems, accounts receivable systems, accounts payable systems and general ledger systems. Shannon's contribution to the Digital Revolution set the foundation for transition from analogue to digital processing of information.

Steve Jobs was born in 1955 and he lived until 2011. He was an American inventor and entrepreneur who, along with his partner Steve Wozniak, co-founded Apple Computer.[24] Jobs and Wozniak started Apple Computer in the Jobs family garage. Together, they revolutionized the computer industry by making personal computers that were cheaper, faster, portable and more user-friendly than mainframe computers. Jobs did the marketing, while Wozniak did the design. Jobs had neither software not hardware skills, but he had effective branding campaigns and stylish designs that appealed to consumers who wanted computing power without having to write computer code. Apple I and Apple II were so successful, they carved out a large share of the personal computer market. The use of the mouse to point-and-click reduced the need to communicate with the computer by writing computer code. Later products, such as iPod and iPhone, revolutionized the industry by providing access to informational resources far beyond those available on hand-held devices at the time. The iPod makes a wide variety of music available. The iPhone is not just a telephone with a texting feature. It is also an alarm, a GPS street locator, with e-mail access and it offers access to the Internet. In addition, there is access to an app store from which applications can be downloaded to perform many logical tasks that used to be performed by humans.

Michael Irwin Jordon is an American who was born in 1956 and is still alive at the time of writing (2019). Having earned degrees in psychology and mathematics, he combined his knowledge of those two subjects to build a career for himself as a computer scientist.[25] His expertise is in Artificial Intelligence (AI), with specialty in Machine Learning (ML). Artificial Intelligence is the science of using computers to mimic human intelligence, by executing code for a process that involves logical thought. Machine Learning is a subset of Artificial Intelligence, in which computer systems learn from data and improve their learning automatically, without being explicitly programmed. In the 1980s, Jordon developed cognitive models for Machine Learning. That involves writing algorithms so that they learn on their own and improve their learning by iterative processes so as to evolve as new data becomes available. The iterative learning does not require any human intervention. In the 1990's, the Artificial Intelligence community started to combine Machine Learning with Data Mining. Computer scientists develop programs that learn by extracting patterns from large amounts of data and creating new knowledge from the results. The data are usually collected from transactions by people or data recorded about people. In 2010, Jordon was awarded the Fellow of the Association for Computing Machinery. He got that award for his work in the theory and application of Machine Learning.

Before the Digital Revolution, our conviction was that humans think; computers do not. The inventions of Shannon, Jobs and Jordon showed us that there is a discrepancy between our conviction and the reality in the external world. Computers have the ability to perform many of the tasks that only thinking humans used to perform. Computers do perform tasks that require logical skills. Mainframe computers are able to execute coded instructions about tasks that humans perform. Personal computers allow users to download apps to perform logical tasks without the users needing to write computer code. Machine Learning learns from data without human intervention.

Throughout the Digital Revolution, humankind discovered new ideas and integrated the emerging content into a coherent new frame of reference for relating to the external world. The activity among the components of the psyche, as well as the feedback loops between psyche and external world resulted in the emergence of machines that perform logical tasks and learn without human involvement. The ideas and images that emerged from the unconscious helped us to generate new digital ways of relating to the external world. Our conviction that only humans have the intelligence to perform logical tasks was an illusion. Not only do computers mimic areas of human intelligence, they do it faster and more reliably than we do. Computer systems are constantly replacing humans in areas of Artificial Intelligence.

Are humans searching for the origin of our psychological projection? Psychology became a discipline in the early 1900's. There have been publications about psychological projection, but it is not yet in the mainstream of human knowledge. Those humans who are familiar with the concept of psychological projection may have searched for the origin of our conviction. They may well have found the origin to be in our psyche. Others may be wondering why there could be such a discrepancy between our conviction and the external world. Still other humans may have moved on to embrace the new technology without bothering to puzzle over any discrepancy. A retraction may have been accomplished by those who pay attention to psychological projections. Others who do not have an explicit knowledge of psychological projection may have retracted by an intuition of psychological growth. We are still in the Digital Revolution. There is time for retraction. Of course, there may be some who prefer to continue believing that computers are incapable of logical thinking. Fortunately, it is possible for humanity to take advantage of new knowledge emerging from the human psyche without us having any comprehension of the psychological projection itself.

The overall outcome of the Digital Revolution is yet to be determined. So far there has been an emergence from the psyche of digital knowledge and practices that enable us to improve the quality of our lives. We have Artificial Intelligence to secure our homes, to detect fraud in bank transactions and to diagnose medical illnesses. By the use of ideas that emerge from our psyche, we created a world in which machines do a lot of the logical thinking that we used for mundane activities. This enables us to establish a digital society where we can place our confidence in Artificial Intelligence to take over digital knowledge, while we prepare for acquiring higher level capabilities in the next revolution. Technology is evolving in tandem with the psyche. As the Digital Revolution progresses, we have been offloading new knowledge to tools and methods that support our lifestyles in a digital society.

The supporters of Singularity may view the Digital Revolution as that milestone in history where technology enslaves or exterminates human. To some extent, enslavement and extermination are already happening. Some humans subordinate their intelligence to technology by their addiction to game videos. Others are so attached to social media that they spend more time in the virtual world than the real world. Those can be regarded as forms of enslavement. Extermination occurs when people are victims of disasters involving sophisticated Artificial Intelligence, for example, airline disasters caused by malfunction of cockpit technology. However, technology does not appear to be outsmarting humans in the general sense. Machine Learning is bypassing humans in the creation of new knowledge. Machines can learn from Big Data without human cooperation. Machines do acquire knowledge by mining data instead of depending on humans to acquire the knowledge, then offload it to technology. What prevents technology from matching human intelligence is that while Machine Learning can acquire rational content knowledge, so far it cannot extract the non-rational elements such as motivation, emotion or intent. Those who promote Singularity understand that the human psyche includes non-rational aspects such as consciousness and a mind. However, they are inclined to focus on what is tangible: the brain. So far, I have not seen any credible explanation that reverse engineering of the brain will imbue technology with human intelligence. Singularity advocates need to address consciousness, mind, intention, emotion and motivation, before claiming that Artificial Intelligence can surpass human intelligence.

The notion of a Singularity is not new. Others have predicted dates in the future when there will be a turning point in the evolution of humanity. In the next chapter, I describe the biographies and predictions made by two scientists. Both predict the possibility of human extinction.

NOTES
1. See "Psychological Types" by Carl Jung. The Collected Works, Volume 6, Paragraphs 577 – 613.
2. See "The Ever-Present Origin" by Jean Gebser.
3. See "The Nature of Technology" by W. Brian Arthur.
4. See "Archetypes and the Collective Unconscious" by Carl Jung. Collected Works, Volume 9, Paragraphs 6 - 7.

5. See stages of projection on pages 113 – 114 of "A Critical Dictionary of Jungian Analysis" by Andrew Samuels, Bani Shorter and Fred Plant.

6. See "Projection and Re-Collection in Jungian psychology" by Marie Louise Von Franz (Open Court Publishing Company, 1978)

7. See a description of Chaos Theory https://fractalfoundation.org/resources/what-is-chaos-theory/

8. See information about complex systems that have a surprise-generating mechanism known as emergence: (https://www.britannica.com/science/complexity-scientific-theory).

9. See a description of the myth about Demeter, Persephone and Hades at web site https://www.windows2universe.org/mythology/persephone_seasons.html&edu=high

10. See information about "De Revolutionibus Orbium Coelestium" ("On the Revolutions of Heavenly Spheres") at the web site https://www.space.com/15684-nicolaus-copernicus.html.

11. See information about the work that Georg Rheticus did to help Copernicus to publish his work on heliocentricity at web site https://www.maa.org/press/maa-reviews/the-first-copernican-georg-joachim-rheticus-and-the-rise-of-the-copernican-revolution

12. See "Rheticus and Copernicus" by John H. Lienhard for information about how the Lutheran theologian influenced the publication of Copernicus' work.

13. See information about the Inquisition's response to Galilei's work "Dialogue Concerning the Two Chief World Systems" at web site http://www.thecosmos.website/Reference/Galileo-Galilei.html

14. See information about the first person to build a steam-powered automobile at web site https://www.history.com/news/who-built-the-first-automobile.

15. See information about the first gasoline-powered automobile at web site https://www.reference.com/history/first-automobile-invented-381896650830f4f1

16. See information about the first mass-produced automobile at web site https://www.history.com/this-day-in-history/fords-assembly-line-starts-rolling.

17. See information about Nicolas-Joseph Cugnot's steam-powered automobile at web site (www.britannica.com).

18. See information about Karl Benz's gasoline-powered, three-wheeled Motorwagen. (www.britannica.com).

19. See information about how Bertha Benz demonstrate the practical use of the Motorwagen at web site https://www.mylifetime.com/she-did-that/august-5-1888-bertha-benz-took-the-first-documented-road-trip-in-an-automobile

20. See information about Henry Ford, the American automobile manufacturer at web site (www.biography.com).

21. See information about psychological projection as defined by Sigmund Freud at web site (www.britannica.com).

22. See information about Claude Shannon's communication model for digital computers at web site (www.britannica.com).

23. See information about computers built using electronic relays and switches. (www.scientificamerican.com).

24. See information about Steve Jobs and Steve Wozniak, co-founders of Apple Computer. (www.biography.com).

25. See information about Michael Irwin Jordon, the AI expert at web site (https://www2.eecs.berkeley.edu/Faculty/Homepages/jordan.html).

Chapter 9

Predictions Based on Physical Sciences

Singularity is not a new topic in the discussion of evolution. Many scientists have offered predictions about major disruptions and turning points in the evolution of life on planet earth. Two of these futurists are scientists who never met, but have written about the same idea: an emerging superintelligence at some point in human evolution. These scientists are from different professions, different cultures and different centuries. Nevertheless, they have both predicted a paradigm shift in the future when technology's impact on the human mind will bring about a major change in the way humans function. They are Pierre Teilhard de Chardin (1881 – 1955) and Raymond Kurzweil (1948 – present). Chardin and Kurzweil have much in common. Both have ideas considered to be outside the mainstream scientific thinking of their day. Both have extensive credentials in scientific domains. Both use their scientific background to bolster their arguments about the coming emergence of a superintelligence, without taking psychology into account. They both acknowledge that consciousness plays a role in the upcoming paradigm shift, but they do not include the human psyche as a factor in their predictions. Although Chardin does use the word "consciousness" in his writing, he does not appear to be using the word in reference to a component of the psyche that exists in opposition to, and in relationship with unconsciousness. Instead, he seems to use the word as a synonym for thought, awareness or wakefulness. Chardin referred to the emergence as the Omega Point; Kurzweil calls it the Technological Singularity. Their predictions are based on the topics on which their careers and interests are grounded. Chardin was a paleontologist and a Jesuit priest. Kurzweil is a computer scientist and a seeker of immortality, who takes vitamin supplements and intravenous longevity treatments in an effort to live forever.[1]

Chardin and Kurzweil both regard technology as a driver in the emergence of a superintelligence. Both scientists have had success in the area of prediction. Decades before the Internet came into existence, Chardin predicted a net encircling the earth as a means of communication among humans. The Internet serves that function now. Kurzweil predicted that by the year 2000, computers would beat the best human chess players. In 1997, IBM's Deep Blue computer beat Gary Kasparov, who was the World Champion.[2]

In the next two sections, I provide brief biographies that include the predictions of Chardin and Kurzweil.

9.1   Pierre Teilhard de Chardin

Pierre Teilhard de Chardin was born in 1881 in Auvergne, France of Catholic parents. He studied mathematics and philosophy. He later attended the University of Paris, where he studied three subjects in natural sciences: botany, zoology and geology. Between 1912 and 1914, he worked in the paleontology Laboratory of National Musee d'Histoire Naturelle in Paris.[3]

In 1914 he served as a stretcher-bearer in World War I. On his return from the war, he entered the Jesuit Order. From 1920 through 1922, he lectured in geology at the Catholic Institute of Paris, while he obtained a Doctorate degree in natural sciences. His publications about evolution and cosmic life came to the attention of the Catholic authorities. In the mid-1920's, after characterizing Chardin's written work as incompatible with Catholic orthodoxy, the Catholic Church relieved him of his lecturing duties at the Catholic Institute of Paris. The Catholic authorities assigned him to do paleontology research in China.[4] In 1923 he travelled to China, where he lived and conducted paleontological research for most of the next twenty years. In 1951, he relocated to the United States. Chardin died in 1955 in New York.[5] Due to conflicts with Catholic Church, his writings were published posthumously. "The Phenomenon of Man" was published in 1955. That is the book in which Chardin offers an account of serial evolution from inanimate matter to Homo Sapiens. He identifies the Omega Point as a turning point in evolution, when human thought will form a sheath encircling the earth. It is a scientific speculation, shored up by a spiritual belief. He saw the curve of evolution as a trend towards superintelligence, greater complexity and increased communication, all guided by a spiritual direction. After the Omega Point, Chardin speculates that the superintelligence which emerges will be a new kind of humanity whose knowledge will encircle the globe in a sheath of communication, while humans will have the option to relocate from planet Earth and to colonize other planets in the universe.[6]

Chardin identified a sequence of spheres that show increasing "complexification" in the sense that each sphere has more evolutionary sophistication than the preceding sphere.[7] The spheres range in increasing evolutionary sophistication from the Geosphere to the Noosphere.

- Geosphere: a stage in evolution when the earth is only populated by molecules of inanimate matter such as rocks and minerals.
- Biosphere: a stage in evolution when the earth's population has grown to include single-cell organisms such as bacteria, fungi and amoeba.
- Atmosphere: a stage in evolution when the earth's population has grown to include creatures of the air, such as birds.

- Noosphere: a stage in evolution when the earth's population has grown to include humans capable of thinking, and whose thought encircles the earth in a "Planetary Mind".

Chardin's prediction is that the sequence of spheres in the "curve of evolution" lead toward a point in human evolution when technology enables a network of human connectivity to encircle the globe, resulting in the Omega Point, with the possible extinction of humanity. Although his prediction is grounded in his career as a paleontologist, it is very similar to Kurzweil's prediction.

9.2   Raymond Kurzweil

Raymond Kurzweil was born in 1948 in New York, United States and he is still alive at the time of writing (2019).[8] He has Jewish immigrant parents, while he is an agnostic.[9] His parents exposed him to various religions and encouraged his interest in science. When he was a young boy, Kurzweil distinguished himself as a computer programmer and an inventor. He wrote his first computer program when he was fifteen years old. In 1965, he appeared on the CBS TV show "I've Got A Secret" where he played piano music created by a computer he had built.[10]

Kurzweil studied at the Massachusetts Institute of Technology, where he earned a bachelor's degree in computer science and literature in 1970.[11] He invented a number of technical products and founded companies to produce them. He founded the company Kurzweil Computer Products in 1974. The company used a computer program to perform optical character recognition, which made it a reading machine for the blind.[12] Inspired by Stevie Wonder, Kurzweil founded Kurzweil Music Systems in 1984 to produce music synthesizers that duplicate the sound of real musical instruments.[13] In 1987, he established a company named Kurzweil Applied Intelligence to perform computer speech recognition.[14] Later, Kurzweil established the company Kurzweil Cybernetic Poet, which creates poetry automatically. After selling each company that he established, he would remain as a consultant for a period of time before moving on to his next invention. In 2012, Google hired Kurzweil to work on Machine Learning and Natural Language Processing.[15]

During the 1990's, Kurzweil established Kurzweil AI.net, which is a web site for sharing resources among the technology elite.[16] In 1999, he was awarded the National Medal of Technology, which was presented to him by president Bill Clinton at the White House.[17] The Wall Street Journal characterized Kurzweil as a "restless genius" while Forbes magazine called him "the ultimate thinking machine" and PBS featured him as one of the sixteen "revolutionaries who made America".[18]

Kurzweil has published several books. In 1990, he published his first book "The Age of Spiritual Machines" which explains that the rate of change in a variety of evolutionary systems tends to increase exponentially. In 2005, he published "The Singularity Is Near" which became a best seller. That is the book in which he made his famous prediction that he anticipates Technological Singularity by the middle of the twenty first century. The Singularity.com web site posted a review of the book. Here is an excerpt from the review:

"In the new world, there will be no clear distinction between human and machine, real reality and virtual reality. We will be able to assume different bodies and take on a range of personae at will. In practical terms, human aging will be reversed; pollution will be stopped; world hunger and poverty will be solved. Nanotechnology will make it possible to create virtually any physical product using inexpensive information processes and will ultimately turn even death into a soluble problem."[19]

Another book by Kurzweil was the 2012 publication of "How to Create a Mind: The Secret of Human Thought Revealed" which explains his idea that brain emulation can lead to artificial superintelligence. Although Kurzweil calls himself an agnostic, the titles of his books seems prone to religious references. The title of the book "The Singularity Is Near" is a Biblical reference to the second coming of Christ. The titles of his other books include the words "Spiritual" and "Transcendent" leaving the reader with the idea they are intended to be inspirational. Examples of such titles are "The Age of Spiritual Machines" and "Transcendent Man".

As a member of the Alcor Life Extension Foundation, Kurzweil has an arrangement that, in case of his death, his body will be frozen with chemicals to preserve his body so that he can be resuscitated at a future point in time.[20] After the event of Technological Singularity, Kurzweil speculates that the emerging superintelligence will be Artificial Intelligence, which will take over control of planet Earth by enslaving or exterminating humanity.

Chardin and Kurzweil both recognize that consciousness plays a role in human evolution, but neither takes psychological principles into account to support their ideas. In the next chapter, I explain my predictions supported by psychological principles.

NOTES

1. See Kurzweil's outlook on immortality
   http://longevityfacts.com/futurist-ray-kurzweil-takes-100-pills-daily-to-live-forever/
2. See IBM's Deep Blue chess championship
   https://futurism.com/ray-kurzweils-most-exciting-predictions-about-the-future-of-humanity/
3. See Chardin's work in paleontology at the National Musee d'Histoire Naturelle in Paris
   http://people.biology.ucsd.edu/bshanks/hosting/work/presentations/symSysHonorsThesis/talkSlides/wikipediaChardin.html
4. See Chardin's relocation to China at the instruction of the Catholic Church
   http://people.biology.ucsd.edu/bshanks/hosting/work/presentations/symSysHonorsThesis/talkSlides/wikipediaChardin.html

5. See Chardin's death in New York http://www.osservatoreromano.va/en/news/teilhard-de-chardin-and-women

6. See description of Omega Point https://www.vice.com/en_us/article/nz7z7q/the-priest-who-believed-in-god-and-the-singularity-pierre-teilhard-de-chardin

7. See explanation of "complexification" https://teilhard.com/2013/08/13/the-noosphere-part-i-teilhard-de-chardins-vision/

8. See Kurzweil's early life in New York https://www.famousinventors.org/ray-kurzweil

9. See reference to Kurzweil's agnostic outlook https://www.famousinventors.org/ray-kurzweil

10. See reference to Kurzweil on CBS TV https://www.c-span.org/video/?194500-1/depth-ray-kurzweil

11. See Kurzweil's accomplishments at Massachusetts Institute of Technology https://www.britannica.com/biography/Raymond-Kurzweil

12. See reference to Kurzweil's reading machine for the blind https://www.britannica.com/biography/Raymond-Kurzweil

13. See Kurzweil's work on music synthesizers https://www.famousinventors.org/ray-kurzweil

14. See Kurzweil's computer speech recognition http://singularity.com/fullbiography.html

15. See Google hires Kurzweil to work on NLP https://www.cnet.com/news/ray-kurzweil-joins-google-as-director-of-engineering/

16. See Kurzweil's AI.net web site https://www.kurzweilai.net/

17. See reference to Kurzweil being awarded the National Medal of Technology https://www.biographies.net/people/en/raymond_kurzweil

18. See reference to Kurzweil being featured on PBS https://www.biographies.net/people/en/raymond_kurzweil

19. See Kurzweil's characterization of death as a soluble problem
    http://singularity.com/aboutthebook.html
20. See Kurzweil's plan to preserve his body, in case he dies
    https://future.fandom.com/wiki/Alcor_Life_Extension_Foundation

Chapter 10

Predictions Based on Psychological Principles

During the lifetimes of Pierre Teilhard de Chardin and Raymond Kurzweil, there has been a growing focus of conscious attention on the impact of technology on the evolution of humans. Chardin and Kurzweil both believe that consciousness plays a role in evolution, but neither takes psychological principles into account to support their predictions. Chardin's Omega Point is a scientific speculation based on his natural sciences expertise, and it is influenced by his belief that evolution is charged with divine direction. Kurzweil's Technological Singularity is a scientific speculation based on Moore's Law applied to computer science, and it is influenced by his belief that technology will mechanize human bodies in a quest for immortality. Neither Chardin nor Kurzweil takes the human psyche into account in their predictions about humans. I will use psychological principles to support my prediction that the current Digital Revolution will give way to a Psychological Revolution.

In predicting a Psychological Revolution, I describe a re-emergence of Analytical Psychology and its enhancement by Artificial Intelligence. When Carl Jung introduced Analytical Psychology in the twentieth century, it did not acquire mainstream acceptance. The failure to gain mainstream acceptance resulted from a number of contributing factors. Analytical Psychology was regarded as a collection of intriguing concepts that could not be proven or falsified, and was therefore disregarded for not being scientific. The appearance of a Jungian cult on the part of Jung's followers undermined their effort to preserve Analytical Psychology as a respectable discipline. The New Age movement's adoption of Analytical Psychology further relegated it to the non-scientific heap. In spite of those disadvantages, Analytical Psychological survived for more than a hundred years; but it is still in search of a scientific footing. The discipline of Analytical Psychology has been kept alive by practitioners trained at institutes such as the C. G. Jung Institute in Zurich. In addition, followers of Jung have written books to clarify and amplify his works. Practitioners have added to the body of knowledge about Analytical Psychology by writing books about their research and case studies from their clinical practice.

Why has Analytical Psychology failed to gain a scientific footing? One explanation could be that Analytical Psychology does not need a scientific footing, because it is unique and has its own footing. If that is so, Analytical Psychology has yet to articulate its own footing. Another explanation could be that the tools necessary for giving Analytical Psychology a scientific footing had not yet been invented. Yet another explanation could be that historically, intuition often precedes articulation of a scientific footing. After all, there were many astronomers who had intuitions about heliocentricity decades before Copernicus' hypothesis articulated a scientific footing. The hypothesis preceded the telescopes crafted by Galileo to provide the empirical evidence for a scientific footing of heliocentricity. So, the intuition preceded the hypothesis, and the hypothesis preceded the evidence. So too, Jung may have provided the intuitions about the psyche for a later generation to mount on a scientific footing. I believe Artificial Intelligence has a mix of tools that can enable a scientific footing for what Jung intuited.

To support my prediction of an upcoming Psychological Revolution, I apply principles of Analytical Psychology. One is Carl Jung's principle of differentiating the unconscious portion of the psyche into the personal unconscious and the collective unconscious.[1] The content of the personal unconscious was once conscious, but was repressed because it conflicted with the self-image of the individual. The collective unconscious is a reservoir of ancestral heritage that belongs to all humanity. My prediction also relies on Jung's principle that the unconscious part of the psyche is populated by archetypes, which are organizers of ideas in the psyche and are the underlying organizers of matter in the external world.[2] Archetypes have a dual nature; they exist partly in the psyche and partly in the external world. To explain the shift in paradigm from the Digital Revolution to the Psychological Revolution, I apply the principle of enantiodromia, as defined by Jung.[3] This principle states that when conscious attention is dominated by focus on an extreme, an equally powerful counter position builds up in unconsciousness. When the counter position emerges, it first obstructs conscious performance, then it breaks through conscious control. In addition, I apply Jung's principle of individuation which is a process of differentiation in which individuals become psychologically distinct from other human beings.[4] I use the principle of psychological projection to support my prediction of a Psychological Revolution by explaining that projections are a normal occurrence in psychological growth. Carl Jung provides the definition of one type of psychological projection. It states that this is a projection from the personal unconscious or the unacceptable part of the personality.[5] Marie-Louise Von Franz extends that definition to include projection from the collective unconscious. She states "wherever known reality stops, where we touch the unknown, there we project an archetypal image." [6] This is the type of projection that I use to explain the evolution of humanity during historical revolutions. I further explain that while the casting of a projection is beyond the reach of rationality, the withdrawal and integration of the projection are the means by which we achieve psychological growth. Those are the principles of Analytical Psychology that I use to support my prediction that the current Digital Revolution will be followed by a Psychological Revolution.

To explain the paradigm shift from the Digital Revolution to the Psychological Revolution, I use the principle of enantiodromia. The principle of enantiodromia supports my prediction that the extreme view in society's conscious attention, during the Digital Revolution, will precipitate the emergence of an opposite view from the unconscious psyche. In an earlier chapter, I provided a detailed explanation of enantiodromia, which indicates that when our conscious attention on a given topic approaches an extreme, it triggers the emergence of an opposing topic from the unconscious part of our psyche. The content that emerges from unconsciousness will be emergent phenomena and therefore not derivable from components of the Digital Age. The Psychological Revolution will not be the logical effect of any cause in the Digital Age. I predict that the emergent phenomena will be about how our psyche interacts with the external world in a collective way, as well as how we manage the interior functions of our psyche. I see Technological Singularity as an approaching extreme in our conscious attention. I predict that when it reaches an extreme, there will be a paradigm shift from the Digital Revolution to a Psychological Revolution. I believe that the Artificial Intelligence tools developed in the Digital Revolution can help Analytical Psychology to acquire a scientific grounding.

Since Jung outlined his Analytical Psychology, there have been some steps toward giving his psychology a scientific footing. There are tests based on Analytical Psychology, for example, the Word Association Test, the Myers-Briggs Type Indicator (MBTI), the Jungian Archetype Test, the Jungian Type Index (JTI), the Jung Typology Profiler for Workplace (JTPW), and Mapping the Organizational Psyche. The personality tests have been found useful for people to gain a better understanding of themselves, and for team building exercises. Some employers use the tests for hiring purposes. It is possible to train workers to perform specific jobs, but not possible to train workers in the personality for the job. So, it is useful to be able to select employees who have personalities that fit the job. The tests have also been useful in the selection of leadership styles. While some find these tests useful, others question the validity of them. I believe that Artificial Intelligence offers a mix of tools that can be used to evaluate the validity of the personality tests.

There are psychologists who have been working to render psychology a scientific discipline. These are mainly the cognitive psychologists. Cognitive Psychology became a discipline in the second half of the nineteenth century, when Ulric Neisser published his ideas about the topic. The focus is on cognitive processes involve sensory input, stimuli and behavioral changes that are measurable. That is certainly a bold move in putting psychology on a scientific footing. However, it is not enough. In my opinion, cognitive psychologists do not pay attention to the psyche as a whole. They do not pay attention to the unconscious influences that impact cognitive activities. I believe that Analytical Psychology has a broader scope, that is more representative of the human psyche.

In the next two sections, I explain my predictions about an "Individuating Mind" and a "Consolidating Mind" that I expect in the Psychological Revolution.

10.1   Predictions at the Personal Level: the "Individuating Mind"

At the personal level, my prediction is that the psyche will have an expanded role, for those who choose to explore the psyche. I use the expression "individuating mind" to mean a mind that deliberately manages psychological growth from the vantage point of continuously acquiring knowledge about the psyche. An individuating mind has these characteristics:

- The individuating mind has a psychological literacy about the psyche that makes it possible to intentionally manage the individuation process. This includes the knowledge that the individuation process covers a number of different activities throughout the course of a life: the emergence of a person's psyche from undifferentiated unconsciousness, the integration of components of the psyche, and continuous effort to maintain balance between consciousness and unconsciousness.

- The individuating mind has the power to take action that bridges life's events to relevant archetypes in the unconscious psyche. This action depends on familiarity with the Jungian principle that the collective unconscious is populated by archetypes which can be activated by events in the external world.

- The individuating mind has the introspection to discern psychological projections, knowing the difference between a projection that is a defense mechanism (projection from the personal unconscious) and a projection that is an opportunity for psychological growth (projection from the collective unconscious).

- The individuating mind has the self-awareness to distinguish between its dominant and latent functions of the psyche, as well as the drive to develop the latent functions. This entails a knowledge of the Jungian Typology.

- The individuating mind has a proficiency in the interpretation of dreams, as a way of understanding communications from the unconscious part of the psyche.

- The individuating mind has an awareness that it feeds knowledge and experience to the "consolidating mind" and therefore affects the future of humanity, for better or worse.

Just as reading, writing and arithmetic are considered essential to the development of children, so too psychological literacy will become essential to maturation of adults. The progress of a life is shaped by the way we manage our individuation. Most of us manage our individuation in a psychological fog of how we experience life, coupled with what little we understand of our psyche. I predict that during the Psychological Revolution, we will allocate resources for comprehension of the psyche and the application of the acquired knowledge to our interactions with the people and institutions in the external world. We will be able to recognize our psychological projections as involuntary occurrences, that are part of normal psychological growth. We will also be able to discern the relevant archetypes activated during projections.

Most of the time, most of us are unaware that we interact with others through the mechanism of psychological projections. When we become aware of our projections, it is no easy matter to withdraw the projections. The withdrawal of a projection cannot be accomplished by rationality alone. I predict that we will shift our focus from reliance on rationality and learn more about how to grow psychologically by identifying, withdrawing and integrating our psychological projections. We will come to see how our psychological projections have been skewing our rationality, unknown to us while we were narrowly focused on being strictly rational.

In managing our individuation, we currently make on-the-spot decisions limited by information that is available to us. What is available to us includes a limited recollection of the events of our past, an awareness of where we are and a sense of what we want to achieve in the future. We often do not notice that we tend to repeat undesirable behaviors. When we do notice, we may not know how to effect change in our lives. I predict that in the Psychological Revolution, individuals will cultivate better knowledge of archetypes, partly to better interact with others, but also to understand our interiority. With the knowledge of the archetypes that are influencing our lives, individuals will be able to better manage their individuation for enhancing psychological growth. Individuals will have the knowledge and the power of action to make the connection between events in their lives and the activated archetypes in the unconscious psyche. People will more deliberately move along the path of individuation to progress their psychological growth. As people improve their psychological capabilities, their individuating minds will continuously inform the consolidating mind.

The individuating minds will have the support of artificially intelligent tools to aid them on the path of individuation. Already, there are apps that provide psychological services. Most of the existing psychological apps are devoted to recovery from mental illness. In the future, I predict there will be more psychological apps to promote and sustain psychological health. People will take control of their psychological health by using interactive software tools. The following are examples of a trend in Artificial Intelligence currently being used to support patients and psychologists in management of mental illness:

- "PTSD Coach" is an app that was created by the Department of Veteran Affairs for Post-Traumatic Stress Disorder. The app provides information about the disorder and makes suggestions on how to cope with the disorder. It is not intended to be a substitute for professional help.[7]
- "Schizophrenia Apps" were developed by psychiatry professor Danielle Shlosser to help schizophrenic patients cope with a tendency to isolation. It enables patients to maintain contacts with their peers and follow through on goals to improve their condition.[8]
- "APA Journals" is the official app produced by the American Psychology Association (APA), as a portal that allows access to research about psychology.[9]
- "DMS 5 Diagnostic Criteria Mobile App" was built to help practitioners, researchers, students and patients to access diagnostic criteria and codes for their own purposes.[10]
- "PsycExplorer" is an app that was developed by a psychology professor for sharing latest news, research and current trends in psychology.[11]
- "iCouch CBT" is an app that was built around methods of cognitive behavior therapy. Professionals use it to help patients

address past memories and modify behaviors to lead healthier lives.[12]

- "iCounselor: OCD" is an app that is not intended to replace professionals, but it does offer help to curb the obsessive-compulsive disorder.[13]

The following are examples of Artificial Intelligence that support psychologists in diagnosis and treatment of mental illness:

- "Ellie" is an early version of an artificially intelligent therapist, but is not intended as a substitute for a practitioner. Built to treat veterans suffering from Post-Traumatic Stress Disorder (PTSD), Ellie can analyze facial expressions, head gestures, eye movement and voice patterns to detect indicators associated with depression and PTSD.[14]
- "Embrace" is an artificially intelligent computer system developed for treatment of epilepsy. Developed by Empatica at the Massachusetts Institute of Technology, this artificially intelligent computer system uses personalized medicine, seizure management and drug discovery to determine treatments for epilepsy.[15]

Individuals are already making use of apps and artificially intelligent systems that embody psychological knowledge. Many people are more comfortable providing their personal information to a computer than a human psychologist. A computer is not judgmental. A computer does not require a client to be in an office at a specific time or day to provide personal information. When using a computer, a patient does not experience self-restraint due to shame. Computers can store knowledge about psychology and process natural language faster than psychologists. Computers can detect patterns faster and more reliably than psychologists. I expect that artificially intelligent systems can generate some treatments for maintaining psychological health and overcoming mental illness faster and more effectively than psychologists. Unlike human psychologists, computers do not go on vacation, and do not end a session because an hour has passed.

Humans will come to realize that our survival of the disruption of the Digital Revolution depends on getting to know our psyche better than artificially intelligent systems know our psyche. That is because developers of artificially intelligent systems have a variety of motivations, which cover the spectrum from noble and high-minded, to downright predatory. I predict that during the upcoming Psychological Revolution, psychologists will play a significant role in helping laypeople to discern and develop latent functions of the psyche: thinking, sensing, feeling and intuiting. Psychologists will pay as much attention to maintaining mental health as recovery from mental illness. Laypeople will become proficient at interpreting dreams, to promote psychological growth. The general population will understand the psychological growth that can be obtained by learning to detect and withdraw psychological projections. Knowledge of our psyche will enable us to evolve beyond our current tendency toward rationality-bound capabilities. The knowledge acquired by individuating minds will continuously feed the consolidating mind.

10.2 Predictions at the Collective Level: the "Consolidating Mind"

Analytical Psychology characterizes the psyche as having an unconscious level and a conscious level. The unconscious level is made up of personal and collective components. The conscious level is all personal; it has no collective component. At the time when Carl Jung introduced Analytical Psychology, his structure of the psyche had no collective component at the conscious level. That structure of the psyche was adequate when we humans were striving to establish ourselves as individuals emerging from an unconscious collective existence. In my opinion, we are approaching the next step in our evolution. Individuals used to shape their identity in terms of factors like nation of origin, family's religion and political ideology. Those boundaries are being eroded. On a conscious level, our psyches are coming into an alignment that generates a new component. I use the expression "consolidating mind" to mean the emergence of an ongoing aggregation of "individuating minds" that is supported by various means of connectivity in a global community. Increasingly, individuals are geographically mobile with skills that can be practiced anywhere in the world because of connections via the Internet. Individuals now live in a global community connected by networked systems. An action in one country can impact the entire global community. A financial decision in one country affects the global economy. An outbreak of a disease in one country quickly passes to other countries through global air travel. What used to be national concerns are becoming global concerns: terrorism, climate change, disease control and cybersecurity are examples. The consolidating mind also has connectivity in other respects: the connecting spontaneity that binds fans in tightly contested football game; the connecting motivation that drove the Arab Spring Revolt, and the connecting sense of unity that mystics share. Now that we are all being connected globally, I see our psyche taking on a new dimension. From a collective perspective, I predict that the psyche will take on a conscious role that is inter-connected globally.

The consolidating mind is a collective conscious component of the psyche that I predict will come into existence during the Psychological Revolution. These are the characteristics of the consolidating mind:

- The consolidating mind is fed by the psychological literacy of an aggregation of individuating minds. It uses the knowledge and experience of the individuating minds when taking action.
- The consolidating mind functions as an aggregate of individuating minds acting in concert as a coherent whole, but it does not act continuously. It takes action when events in the external world trigger the activation of archetypes.
- The consolidating mind has a quality similar to the spontaneity of fans at a ball game. Fans have a knowledge of the rules of the game that enables them act with spontaneity. Uproariously, fans

applaud players who score points, and boo referees who mis-apply the rules.

- The consolidating mind has the emotion-toned motivation coupled with the action-orientation of the Arab Spring revolt in 2011. Young, well-educated Arabs expressed their collective, conscious outrage about the dismal economic circumstances in multiple countries including Tunisia, Yemen, Syria, Egypt, Libya, Algeria, Iraq, Jordan, Kuwait, Oman and Morocco. The external event that ignited their outrage was a food vendor who burned himself alive in front of a government building, when he despaired of making a living after being denied a vending license.

- The consolidating mind has a characteristic sense of oneness in common with mystics. The mystical experience is not planned or organized. Although mystics may have difficulty explaining their experience, they express a common reaction. The mystical experience bestows a powerful sense of oneness with all of humanity. During a mystical experience, the mystic feels a sense of unity with the whole universe.

- The consolidating mind transcends geographic boundaries. It can involve large swaths of humanity, without regard for geographic location. Examples are global responses to events such as space exploration, international sports events, scientific discoveries, acts of terrorism, war, changes in global climate and natural disasters.

- The consolidating mind will marshal the efforts of psychologists to promote psychological literacy, create opportunities for psychological growth, and establish curricula for education about sustaining psychological growth.

The coming together of individuating minds is triggered by events that impact large portions of humanity in a global way. We currently see early forms of this response in recovery from natural disasters, for example, earthquakes, hurricanes and terrorist attacks. The disaster triggers a response from people all over the globe: some set up web sites for keeping track of people who are missing; others collect funds via smart phones; still others volunteer their resources to provide hot meals and other basic necessities. All of this voluntary activity occurs without any central organization and without any hierarchy of leadership. Another example of a trigger for the consolidating mind is the outbreak of contagious diseases that threaten global public health such as new strains of influenza, tuberculosis, the Ebola virus and SARS (Severe Acute Respiratory Syndrome). Given the interest of enough individuating minds, the consolidating mind can also trigger a major cultural event, like the formation of political alliance of countries that have shared interests, or a unified economic body for member countries. We currently have such global bodies, but they come about by having several organizations engaged in frictional collaboration with reliance primarily on rationality, over extended periods of time. The consolidating mind will be more effective because it has a global perspective; it will take more humane considerations into account, and function faster when enough individuating minds are aligned.

As the consolidating mind emerges, it will become a global custodian of planet Earth. The consolidating mind will have a shared sense of purpose and will take advantage of the immediacy of technological communication similar to what we saw in the Arab Spring[16] revolt of 2011. At that time, young citizens of Arab countries revolted against the old guard of authoritarian leadership. With a shared sense of purpose, young citizens across Arab countries took to their cell phones for rapid communication of their outrage when the Arab media was not allowed to report the uprising. Their collective outage toppled leaders in countries such as Tunisia, Egypt and Saudi Arabia, while resulting in social reforms in other Arab countries. The consolidating mind will possess multifunctional abilities as in an organization where there is distributed authority and no individual leader. Knowledge is distributed throughout the consolidating mind, which impels action without the labor of rational decision-making or planning, and without the need of hierarchical leadership.

Over the course of the Psychological Revolution, the consolidating mind will come to function like the psychological equivalent of the rationality-bounded organizations such as the United Nations, the North Atlantic Treaty Organization and the European Union. However, the consolidating mind will gradually erode national boundaries in domains of economics, finance, climate, education and ethics, to function as a collective entity. The consolidating mind depends, for better or worse, on knowledge continuously being acquired by the individuating minds.

10.3   Connecting Jungian Intuitives to Artificial Intelligence

What I have noticed about the people drawn to Analytical Psychology is that they are predominantly Intuitives, in the sense of the Typology of functions of the psyche. Jung himself was an Intuitive. We owe a debt of gratitude to Intuitives because they are able to imagine concepts long before anyone can prove them. However, it is unrealistic of us to expect Intuitives to set psychology on a scientific footing. Intuitives are not comfortable with structure. They find procedures restricting of their style. Analytical Psychology did not achieve mainstream acceptance when it was first introduced in the early twentieth century because the Intuitives do not care for the structures and procedures necessary for a scientific footing. The criticism was that Analytical Psychology contained concepts and principles that were largely intuitive; they could not be proven or falsified. I believe we now have technology that can help psychologists to put Analytical Psychology on a scientific footing. The technology is in the mix of tools available in Artificial Intelligence. To achieve a scientific footing for Analytical Psychology, it may be necessary to pair Artificial Intelligence experts (thinking tends to be their dominant function) with Analytical Psychologists (intuition tends to be their dominant function).

Although work on Artificial Intelligence began in the middle of the twentieth century, it did not come into general usage until the 1990's. In the early life of Artificial Intelligence, algorithms were knowledge-driven, in the sense that humans had to write computer programs to offload our knowledge to technology. In the Digital Age, algorithms have become data-driven, meaning that they produce results by mining repositories of Big Data. Big Data are large collections of data accumulated from various human activities, for example, search engine results, financial trading transactions, health care treatments and political activism. Computers analyze large volumes of data and generate new knowledge about the types of human mental activities that produced the Big Data.

Artificial Intelligence uses algorithms to mine Big Data for the purpose of extracting clusters of data, and continuously improve guesses of what the humans were thinking when they produced the data. Artificial Intelligence uses Big Data to extract knowledge from human transactions. Artificial Intelligence has the advantage that it can extract implicit knowledge that humans possess, but are not necessarily able to articulate. If Machine Learning can discover our implicit knowledge, it means that algorithms are learning more about us than we know explicitly about ourselves. By successive extractions of knowledge from Big Data, Machine Learning can also improve its learning about our mental processes in more nuanced ways than we humans improve our learning.

The learning ability of machines is demonstrated by the development of artificially intelligent systems made up of algorithms. Some can interact with humans. For example, Automated Healthcare Systems, is an artificially intelligent system that proposes treatments when given symptoms of diseases. Experts are also developing algorithms to extract implicit knowledge from repositories of Big Data to provide stand-alone services. Here are some examples of the use of artificially intelligent algorithms:

- NETFLIX uses algorithms to create personalized content based on customers' past purchases of movies.[17]
- SNAPCHAT has algorithms for filtering pictures posted to the website.[18]
- OVAL MONEY contains algorithms that offer customers different strategies for saving money when making purchases.[19]
- GOOGLE MAPS has algorithms that help drivers to choose parking spots.[20]

Algorithms are being created from data that has been collected about human activities. The data collection often occurs when we engage in activities that are recorded online, but it is also collected from face-to-face interactions such as visits to the doctor's office. We do not have to develop comprehensive instructions for algorithms. The artificially intelligent algorithms are smart enough to find patterns in Big Data and extract new knowledge about human mental activity without us being aware of how our data are being used.

As the Digital Revolution progresses, I surmise that we will have a new identity defined in terms of the type of learning we leave to machines and the type of learning we do ourselves. We have already offloaded to Artificial Intelligence many tasks that depend on rationality. I believe we will continue to offload rationality-based tasks to technology. Then we will turn our attention to our non-rational capabilities. By "non-rational" I mean psychological capabilities that are foundational to the human psyche, but which have been subordinate to the place of honor we give to rationality.

FIGURE 10.1

TECHNOLOGY CAN HELP ANALYTICAL PSYCHOLOGY GAIN SCIENTIFIC FOOTING

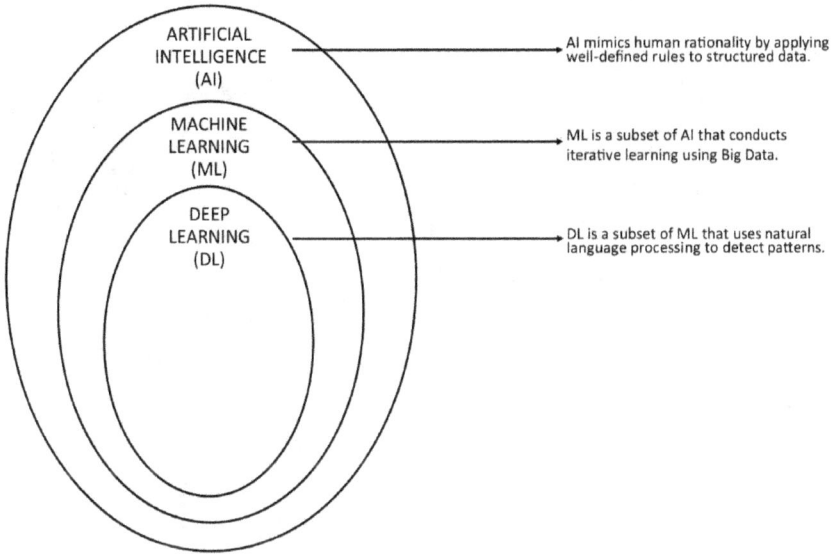

ARTIFICIAL INTELLIGENCE (AI) — AI mimics human rationality by applying well-defined rules to structured data.

MACHINE LEARNING (ML) — ML is a subset of AI that conducts iterative learning using Big Data.

DEEP LEARNING (DL) — DL is a subset of ML that uses natural language processing to detect patterns.

Figure 10.1 shows that technology has much to offer in helping Analytical Psychology to establish a scientific footing. Technology offers three main ways of supporting Analytical Psychology. They are Artificial Intelligence, Machine Learning and Deep Learning, all of which use different methods of acquiring knowledge that can be used for the purpose of providing proof or falsification of psychological principles:

- Artificial Intelligence mimics aspects of human rationality by applying well-defined rules to structured datasets.
- Machine Learning applies iterative learning to accumulate and improve knowledge acquired from continuously updated Big Data.
- Deep Learning uses natural language processing to detect patterns in written or spoken communication.

In the next three sections, I describe ways in which technology can support Analytical Psychology's quest for a scientific footing by use of these three branches of technology.

10.4 Deep Learning: Simulation of Natural Language Processing

Deep Learning is a simulation of aspects of natural language processing. It uses algorithms to emulate human learning by detection of patterns in natural language. It models the type of learning that humans use to detect patterns in language. Deep Learning is potentially useful to Analytical Psychology because it involves the creation of algorithms that can process natural language communication between humans. In "The Symbolic Quest", Edward Whitmont points out that the expression "talking therapy" was used to characterize Analytical Psychology when it was a new discipline, and before it recognized that the verbal interaction needed to be supplemented by a bodily component to be effective.[21] Since then, therapeutic techniques have been expanded to include artistic expression, the role of imagination as well as the awareness of affect and body. Since talking is still a significant action in analysis, it produces quantities of written and spoken language. The processing of natural language is the essence of Deep Learning. Dreams are often a subject of conversation in therapeutic sessions. The reporting of dreams and the interpretation of dreams are communicated in natural language. The conscious part of the psyche, specifically the ego, communicates through concepts expressed in natural language, such as English, Spanish or French. The unconscious part of the psyche communicates through images as expressed in dreams. The interpretation of dreams requires relating images to concepts. Analytical Psychology lends itself to Deep Learning because the interaction between analysts and analysands takes the form of conversations, that is, natural language. Dream interpretation is a good example of the natural language in Analytical Psychology. Some of the content of conversations is about reporting dreams and interpreting the dreams. I think that Deep Learning can play a supporting role in helping to obtain a scientific footing for Analytical Psychology, because natural language processing is applicable to the spoken and written language of dreams, also conversations between analyst and analysand.

Whitmont draws on his experience as a psychologist, to inform us that the dream is purposive.[22] The purpose is for images coming from the unconscious part of the psyche to compensate for one-sidedness in the conscious part of the psyche. Dream images aim to expand consciousness by being allowed to into conscious life. For that reason, the psyche strives for an ongoing relationship between consciousness and unconsciousness. By balancing the dreamer's waking point of view, the dream provides a means for supporting psychological development. The purposiveness of dreams can be found in the layers of communications between analyst and analysand. That is where Deep Learning can be useful. Deep Learning can help Analytical Psychologists to verify or falsify principles based on the structure of the psyche and the purpose of dreams. It can probe the layers of natural language communication to detect patterns in dreams and conversations.

Psychological projections are also purposive. Although the projections are cast involuntarily, the recognition and withdrawal are necessary for psychological growth. The projection has a sequence of stages that entails patterns. Repetitive patterns of thoughts and behaviors and emotional reactions. If these are expressed in natural language, they can be the subject of natural language processing. Projections and dreams have something in common. Dreams sometimes occur in series. A person may experience a series of dreams about the same topic, until the person understands the dream and incorporates its meaning into consciousness. Projections require time for recognition, withdrawal and integration into consciousness. If the stages of the projection are expressed in natural language, the projection can also be the subject of natural language processing in Deep Learning.

See Section 10.1 for examples of technology being used to support patients and psychologists in the diagnosis and management of mental illness.

10.5   Machine Learning: Simulation of Human Learning

Machine Learning is a simulation of aspects of human learning. It is a branch of Artificial Intelligence where algorithms can learn continuously by abstraction from increasingly large sets of data, without human intervention. Machine Learning can be used to produce predictive analyses, based on iterative learning from Big Data about psychological topics. Projections, dreams and traumas all involve a series of activities over a period of time. What Machine Learning has to offer is that it can work with unstructured data to extract clusters of information. Clusters extracted from psychological data may pertain to projections, dreams, inter-generational legacy (e.g., unconscious guilt) and trauma. In every cycle of learning, the datasets are increased by the addition of more data about projections, dreams and trauma. With each iteration of learning, the clusters are updated. Algorithms can use the ever-updated clusters to generate predictions, and verify psychological principles. Another example of the application of Machine Learning to Analytical Psychology could involve accumulating Big Data about inter-generational legacy, as in unconscious guilt or trauma.

Overcoming the guilt and the trauma can extend more than one generation. Recovery from the effects of guilt and trauma involves a series of intrapsychic dynamics over a period of time. Recovery generates data. Inter-generational legacy generates data. Psychologists and researchers could design studies about inter-generational legacy to help give Analytical Psychology a scientific basis. First the psychologists and researchers would decide which psychological principles about inter-generational legacy are to be tested. Then, they would develop algorithms to learn iteratively from data collected about populations of people over the durations of the relevant generations. They would develop algorithms to extract clusters of data about guilt and traumas.

Given enough data and cycles of learning, they could develop other algorithms to process the clusters and produce analyses about the trends in recovery from guilt and trauma. Psychologists would then need to review the analyses to determine which principles are verified and which are falsified. Such an approach would establish a source of continuously, self-improving knowledge bank of validated principles related to guilt and trauma, as defined in Analytical Psychology.

See Section 10.1 for examples of technology being used to support patients and psychologists in the diagnosis and management of mental illness.

10.6   Artificial Intelligence: Simulation of Human Rationality

Artificial Intelligence is a simulation of aspects of human rationality. In this section, I address the type of artificial intelligence that involves computer systems which process decisions in the rational way that humans usually do. In the area of rationality, the value that computer systems add it that they make the decisions faster and more reliably than humans. Artificial Intelligence can be used to produce statistical analyses of data about mental health and mental illness, based on principles in Analytical Psychology. Psychologists have the opportunity to identify which principles would be helpful in putting Analytical Psychology on a scientific footing. Then, they would design studies, collect data, and develop computer systems to produce analyses.

Artificial Intelligence can support psychology by developing computer systems that use proven principles as the basis for selecting treatments for patients. The computer systems would take a number of sources of input into consideration. One sources would be a bank of psychological principles that have been verified. Another source would be new research in psychology. Input to the computer system would be information about the condition of a patient: symptoms, artwork, affect and bodily awareness. The computer system would compare input from the patients with the bank of knowledge about verified principles to produce recommendations of possible treatments. Psychologists would review the recommendations and select a treatment for the particular patient. This approach would relieve psychologists of the laborious aspects of analysis and reduce the time necessary for treatment. In addition, it helps to put Analytical Psychology on a scientific footing because, the work proceeds on the basis of verified psychological principles. Artificial Intelligence is already being used for diagnosis in health care. In addition, there are apps and algorithms that help patients manage schizophrenia, trauma and obsessive-compulsive disorder.

Branches of Artificial Intelligence can be combined to support Analytical Psychology. Deep Learning detects patterns. Machine Learning conducts iterative learning from large repositories of data. Data could include a bank of symbolism in dreams, a bank of knowledge about research on trauma, and a bank of psychological principles that have been verified. Deep Learning algorithms and Machine Learning algorithms can be combined to support Analytical Psychology in dream interpretation and withdrawal of projections. Dream interpretation involves accumulation of knowledge about the dream topic. Projection involves accumulation of knowledge about stages of the projection. Together Deep Learning and Machine Learning can help Analytical Psychology obtain a scientific footing by designing studies to collect data and make interpretations based on psychological principles. Deep Learning is useful in studies that take a non-linear, non-rational approach. Deep Learning uses natural language processing to detect patterns. These could be patterns in dreams and patterns of behavior when archetypes are activated. Such studies would address principles of Analytical Psychology, for the purpose of determining whether the principles are proven or falsified.

See Section 10.1 for examples of Artificial Intelligence being used to support patients and psychologists in the diagnosis and management of mental illness.

10.7 Approach for Applying Artificial Intelligence to Analytical Psychology

My prediction is that Artificial Intelligence is poised to help psychologists obtain a scientific basis for Analytical Psychology. Artificial Intelligence has demonstrated strong capability in the simulation of rational intelligence, as well as simulation of natural language processing. If psychologists are not familiar with Artificial Intelligence, they can be paired with researchers who do have experience with Artificial Intelligence. They could design studies that use natural language communication collected from individuals about dreams, trauma and projections. When the technology produces the results of studies, psychologists would make the final decisions about which principles are verified and which are falsified.

While Artificial Intelligence performs analyses faster and more reliably than humans, Artificial Intelligence does not know what is psychologically correct and does not know the code of ethics by which psychologists function. Below, I offer a set of activities that may help psychologists and Artificial Intelligence experts to jointly set Analytical Psychology on a scientific footing. The performers in these activities are psychologists, researchers and artificially intelligent algorithms.

According to Analytical Psychology, the human psyche has conscious and unconscious components, but consciousness has no direct access to unconsciousness. Humans become aware of unconscious content of the psyche through images in projections, dreams, and imagination. The unconscious content of the psyche is not directly accessible to consciousness, nevertheless, psychological growth depends on deciphering images from the unconscious layer of the psyche.

I propose the following activities as an approach for psychologists to use Artificial Intelligence to acquire a scientific footing, by establishing studies to determine if the principles of Analytical Psychology are verified or falsified.

1. Set a goal of verifying or falsifying an identified principle in Analytical Psychology.
2. Design a scientific study to uses algorithms to prove or falsify the principle.
3. Select participants to be subjects and control groups of the study.
4. Determine what type of data and communications are necessary to prove or falsify the principle.
5. Arrange the collection of repositories of data as well as collection of spoken and written communications.
6. Develop combinations of algorithms to detect patterns, learn iteratively and analyze data.
7. Conduct regular assessments of the psychological status of participants and record results.
8. Capture relevant data about participants:
   a. Natural language descriptions of dreams, trauma, fantasies, projections
   b. Drawn images
   c. Facial expressions
   d. Genetic predispositions
   e. Inter-generational legacy, e.g., guilt, trauma
   f. Observations about the behavior of participants.
9. Execute the algorithms using relevant technology about:
   a. Natural language processing
   b. Iterative learning
   c. Speech recognition
   d. Facial recognition
   e. Digitization of artwork, drawn images
   f. Pattern recognition software.
10. Review results produced by the algorithms and determine whether the psychological principle is proven or falsified:
    a. If proven, add the principle to a bank of proven principles in Analytical Psychology.

b.   If falsified, reject the principle.

c.   If inconclusive, modify the algorithm, or the principle, and test again.

11. Repeat the steps above to accumulate a bank of scientifically verified principles in Analytical Psychology.

Psychologists and researchers will have their own approach depending on the psychological principles being tested. Studies that use Artificial Intelligence to obtain a scientific basis for Analytical Psychology will prepare psychologists for the paradigm shift from Digital Revolution to Psychological Revolution. In the next section, I explain my prediction of the enantiodromia involved in making the paradigm shift from the current Digital Revolution to a Psychological Revolution.

10.8   Prediction of Enantiodromia: From Digital to Psychological Revolution

Artificial Intelligence has experienced periods of expansive growth followed by periods called "winters of AI" when there is little activity due to reduction of funding, because Artificial Intelligence over-promised and failed to deliver. In my opinion, the promotion of Singularity is occurring in a period of growth for Artificial Intelligence. Will the Singularity threat of technology controlling humans be realized by the middle of the twenty first century? I doubt it. I interpret Jung's definition of enantiodromia to mean that Singularity is an extreme, which will precipitate a counter movement in the psyche. I predict that the counter movement will be a paradigm shift from our current Digital Revolution to a Psychological Revolution. The Digital Revolution will be disrupted by bifurcations that destabilize the ascendency of technology in the pursuit of Singularity, and that will initiate a shift in paradigm from extreme focus on the strictly rational to a focus that encompasses other capabilities of the psyche. Other capabilities include creativity and emotion.

In the Digital Revolution, focus of conscious attention is currently on technology and its impact on the evolution of humans. More specifically, the focus is on technology surpassing human intelligence, with the expected consequence that technology will control humans. The thinking is that since technology is capable of recursive learning and humans are not, then technology will become smarter, faster and outstrip human intelligence. That is an extreme view that currently occupies conscious attention. Because conscious attention is preoccupied with such an extreme view, we are beginning to see bifurcations which, in my prediction, will trigger an emergence of a contrary view from the unconscious part of the psyche. According to Analytical Psychology, enantiodromia first inhibits conscious performance, then breaks through conscious control. Variables that are currently bifurcating include intelligence and communications.

Historian Yuval Noah Harari identifies one bifurcation. In "Homo Deus: A Brief History of Tomorrow", he points out that intelligence is being decoupled from consciousness.[23] Humans naturally possess a combination of intelligence and consciousness. Technology is building artificially intelligent systems that demonstrate increasing intelligence, the rational kind of intelligence, but they have no consciousness. One bifurcation is the separation of intelligence into intelligence with consciousness (humans) and intelligence without consciousness (Artificial Intelligence).

Communication is also experiencing a bifurcation. The Internet has brought about a bifurcation of communications. Humans used to engage each other predominantly in face-to-face communications. For example, we used to have participants co-located in conference rooms for business meetings. Now we use Skype or WebEx for online meetings of participants who are geographically disbursed. Dating used to be conducted in-person; now there are online dating systems that offer virtual interaction for casual friendships as well as opportunities to choose lifetime partners. There is also a split in commercial communications. In the past, we would go to brick-and-mortar stores to make purchases; now we shop online. Instead of going into the bank and conducting business with a teller, we can now do online banking. We used to communicate by handwritten letters; now, the Internet enables us to broadcast e-mails. The bifurcation in communication produced online communication and face-to-face communication.

The bifurcations in intelligence and communication both create greater opportunities, but they also have an inhibiting effect on human performance, because they give rise to questions and concerns about how to perform day-to-day activities. Uncertainties inhibit our conscious performance. Are humans useful economic entities or are robots more financially useful? Which functions in the workplace can be transferred from humans to automated equipment without impairing productivity? Which roles need to remain with humans due to the non-rational, sensitive nature of the work to be performed? Where is it more advantageous to assign mission critical responsibilities …to humans or to artificially intelligent systems? When should we augment humans with electronic devices, and when should we substitute a robot for a human? The bifurcations in intelligence and communication create uncertainties that inhibit our conscious performance. The inhibition of conscious performance is a precursor to the emergence of a counter movement from the unconscious psyche.

The breakthrough in conscious control is yet to come. My prediction is that the breakthrough will be triggered by a technological failure, where the recovery cannot be accomplished by technology alone. The technological failure will have a negative impact on large swaths of geography and the lives of many people. An emergence will breakthrough. The nature of emergent phenomenon is that it is novel. It is unique. It is not subject to cause and effect. It is not made up of components of any past event. In short, it cannot be predicted with specificity.

The emergence that I expect from our unconscious psyche is an eruption of a sense of life-supporting interdependence among those affected by the technological failure. A collective neediness that demands attention be paid to the non-rational attributes that hold us together as a global community. A re-ordering of priorities. A sensing that it is as important to live a social life in a global community as it is to be individually successful. A knowing that we are not merely independent individuals forever competing with each other, but a tier in an inter-generational legacy .... a branch of an unseen, but flourishing rhizome.

Technology is fashioned from domains like software, hardware, data, cognitive science, neuroscience, pattern recognition, speech recognition and voice recognition. These domains are all based on rationality. Although technology is promoted as having artificial intelligence that is comparable with human intelligence, the reality is that technology is limited to what we know how to digitize. However, there is more to human intelligence than rationality. There is creativity. Emotion. Intuition. Motivation. Values. Insight. Morality. Ethics. These are dimensions of human intelligence. None of them depends on rationality. Technology has already surpassed humans in skills that are based on rationality. Technology can perform rational skills faster, more accurately and more reliably than humans. However, when we view human intelligence in its multiple dimensions, technology has much ground to cover in catching up. In addition, technology has not acquired a definition of consciousness, which would be essential to producing a technological equivalent. Human intelligence involves the entire psyche, which includes conscious as well as unconscious aspects. Creativity and emotions involve engaging both conscious and unconscious parts of the psyche. Technology has not made progress in determining how to produce the equivalent of consciousness, unconsciousness, or the engagement between psyche and the external world.

In the next section, I explain my predictions about rationality becoming a utility as we shift our focus to non-rational capabilities of the psyche.

10.9 Rationality Becomes Utility & Focus Shifts to Non-Rational Functions of Psyche

In the Digital Revolution, the human emphasis on rational activities involved in cognition will diminish because cognitive capabilities are being offloaded to technology. Cognitive capabilities will be available on demand to humans, as in the provision of a utility. Cognitive capabilities will take the form of artificially intelligent systems, modular "apps" and algorithms, all of which rely on rationality. Having mastered cognition, humans will evolve beyond the focus of honing cognitive skills. Although humans will retain cognitive abilities, it will be easier and faster and more reliable to download apps to perform cognitive tasks. For example, there are already apps to perform activities such as translate languages, calculate mortgages and plan retirements. What we have yet to do is socialized the notion that apps are available as a utility. We have not yet published mechanisms for the public to choose when to invoke an expert system, when to download an app and when to set up an algorithm to conduct Machine Learning. We do not yet have a mechanism for the public to manage technology that embodies cognitive skills as a utility.

I anticipate a paradigm shift that takes us from the Digital Revolution to a Psychological Revolution. The Psychological Revolution will focus attention on promotion of psychological literacy. This involves making the public aware of the relationship between the conscious and unconscious aspects of the psyche, as well as the interaction between psyche and external world. In the twenty first century, the Digital Revolution we will continue offloading human rational capabilities to technology. Although humans will retain the ability to think and act rationally, we will be able to download software faster than we are able to perform rational activities ourselves. Rational activities include logical analyses, calculations and any activities that can be digitized. Software that embodies rational capabilities are already available as "apps" which are modular software applications that can be purchased from an app store or downloaded free of cost from the

Internet. These apps can be executed on a wide variety of platforms including smartphones, personal computers, tablets and laptops. Examples of existing apps are:

- COMMON APP is an undergraduate admission application that the public can use to apply to over 700 colleges and universities in the United States.[24]
- MAIL & CALENDAR is an app that helps users to manage their e-mail and maintain communication with contacts who are important.[25]
- DOULINGO is language-learning software that English speakers use to learn 16 languages by selecting the picture of a word in the foreign language, listening to the pronunciation, then using it to type translations.[26]
- PLEX is an app that organizes photo and video collections.[27]
- MORTGAGE CALCULATOR is an app that real estate website Zillow.com offers home buyers to calculate mortgage payments, when comparing homes for purchase.[28]

The public will be able to choose apps as a utility to perform rational activities for which we no longer need to allocate human resources.

In the upcoming Psychological Revolution, I expect our psyche to function much differently from the present functioning. We will know our psyche much better than we do today. We will need to acquire a more nuanced understanding of how our psyche functions because we will be interacting with algorithms that are capable of learning to mimic aspects of our psyche. Some of those algorithms will be created by well-intentioned researchers, others will be the work of manipulative merchants. We will use the knowledge of our psyche to shift our reliance on rationality to embrace a more multi-dimensional way of being. Technology will continue to progress and Artificial Intelligence will become more sophisticated at providing rationality-based services and devices. We can take advantage of the fact that, as we offload rationality-related tasks to technology, we are freeing our human resources to explore the latent functions of our psyche.

In the next section, I share my observations regarding trends in movies about Artificial Intelligence.

10.10 Movies Reflect Trend in Relationship Between Humans & Artificial Intelligence

In the world of literature, there are stories that stay with us because they resonate with our psyche. Such stories are in books that are re-printed over many years. The stories are played out in theaters repeatedly. We know how the stories end, but we keep going back to see the theatrical performances. That is because they tell us something about our psyche. The Scarlet Letter. Othello. The Great Gatsby. Little Women. The Nutcracker. The Phantom of the Opera. Just as literature finds resonance with our psyche when it reflects our inner tendencies, so too, movies find resonance with our psyche. There are movies about Artificial Intelligence that may have an enduring effect. In my opinion, they give us some insight into our need to understand our psyche better than algorithms do. Because algorithms are capable of learning, they extract information about us from our communications. Over time, they can accumulate knowledge about us. The knowledge that algorithms accumulate about us can be used to serve us or to exploit us.

Earlier movies about Artificial Intelligence such as "Terminator" and "Matrix" involve weapons and warring groups. The plots are strongly related to what is physical and less related to what is psychological. In those movies, the humans win the fights. More recent movies about Artificial Intelligence, such as "HER" and "Ex Machina", involve neither intimidating weapons nor prolonged physical fights. The plots involve a few humans and algorithms engaged in encounters that are primarily psychological. What I find noteworthy is that the humans are the winners in the physical encounters, while the algorithms are the winners in the psychological encounters. When algorithms outsmart the humans, it is not because Artificial Intelligence has surpassed human intelligence, but because the humans do not know themselves psychologically. The algorithms are able to learn iteratively and they accumulate knowledge about the psychology of their human counterparts from conversations. From natural language conversations, algorithms can build up knowledge about human preferences, needs, fears and aspirations. If the humans have a weak sense of identity and lack the wherewithal to be self-reflective, the algorithms come to know the humans better than the humans know their own psyches. The cultural trend of movies about Artificial Intelligence informs us that we need to understand our psyche better than algorithms can accumulate knowledge about our psyche.

1984: Synopsis of movie "Terminator"

The symbolism in the "Terminator" pertains to the fate of humanity versus technology. The plot of the movie contradicts the theme of Greek mythology in which a person cannot change their fate. In Greek mythology, any attempt to change one's fate will only result in tragedy. The movie's theme is that humans do have the opportunity to change their fate. At the start of the movie, the heroine is an ordinary young woman with no special characteristics or accomplishments to her credit. No special aspirations either. Yet, she takes charge of her doomed fate and manages to kill the Terminator, the perfect killing machine. At the end of the movie, the heroine is recording a message for her unborn son while a storm is about to hit. In spite of the danger of the storm, she fearlessly drives toward it. Her actions symbolize humanity's willingness to fight for what is worthwhile, even when life may be at risk and the goal of the fight is different from one's fate.

This movie is set in the year 1984. It begins with a robot named Terminator. He is an assassin who arrives back in time from the year 2029. A human soldier named Kyle is also sent back in time from the year 2029. The Terminator searches the telephone directory for women named Sarah Connor and kills them. The Terminator finds the last

Sarah Connor in a nightclub, but Kyle saves her from the Terminator. Kyle and Sarah steal a car and escape, but the Terminator pursues them.

Kyle explains to Sarah that an artificially intelligent defense network, known as Skynet, will soon become self-aware and create a nuclear disaster. He also explains that Sarah will have a son named John, who will arrange for the survivors of the disaster to get together and lead a protest movement against Skynet. Kyle further explains that Skynet sent the Terminator back in time to kill Sarah before John is born, in order to ensure there is no protest movement. Kyle also explains to Sarah that the Terminator looks human, but is really a robot, that can kill without hesitation because it lacks human feelings.

Subsequently, there are several physical fights between Skynet operatives and humans. In one of those encounters, the Terminator attacks a police station, killing several police officers in its attempt to locate Sarah, who escapes with Kyle's help. They steal a car and hide in a motel. Kyle informs Sarah that he has been in love with her since he saw a picture of her in 2029. They consummate their mutual admiration for each other. As the Terminator continues the hunt for Sarah, there are several ferocious fights between humans and robots. There is also violent interaction between them during a car chase, a motorcycle chase and a truck chase, while Kyle dispenses pipe bombs to keep the Terminator at bay.

The Terminator locates them hiding in factory, where Kyle and Sarah trick the him by activating machinery that crushes him to death. Kyle also dies in the fight. Some months afterwards, Sarah is a pregnant tourist travelling in Mexico, and recording her experience on audio tapes to share with her son John. A local boy snaps a photograph of her and she purchases it. That is the photograph that her son John shows to Kyle in 2029. Sarah has thwarted her fate as an ordinary human and set in motion the event where her unborn son will organize a protest against Skynet, to prevent a nuclear disaster.

1999: Synopsis of movie "The Matrix"

"The Matrix" is a movie in which the behavior of main character depicts the activation of the savior archetype. He lives two separate lives, one as an ordinary mortal, the other as a god with supernatural powers. Given the choice of two pills, he chooses the pill that offers an opportunity to discover the truth. The symbolism is comparable to the biblical story of Eden, when the humans choose knowledge over obedience. Those around him wonder if he is "the One", a long-expected savior who will save humans from enslavement by robots. When he hovers between life and death, he is resurrected by a kiss. Eros revives the hero. At the end of the movie, the hero ascends toward the sky after leaving a telephoned messianic message to the robots that he will show prisoners a world where is anything is possible.

The hero of the story is a computer programmer named Thomas by day, and a hacker named Neo at night. Neo is curious about the phrase "the Matrix" which appears repeatedly in online interactions. A beautiful woman named Trinity informs Neo that Morpheus can explain the meaning of the phrase. She leads Neo to an underworld where he can meet Morpheus. In the underworld, Morpheus' team engages a hostile group of Agents in a brutal battle. The Agents are the gatekeepers of the Matrix and they are responsible for keeping out intruders like Morpheus.

The Matrix came into existence after a war between humans and robots. Most of the human race is being used as a power supply. Their bodies are asleep and enclosed in pods of liquid. Their minds are plugged into the Matrix in a mechanism that draws bioelectric power from their bodies. The Matrix is a virtual world that has been created to shield humans from the truth that they are now slaves. In the Matrix, humans are grown. The dead humans are liquefied and fed to the living intravenously. That cycle provides an ongoing source of energy for the robots.

Morpheus offers Neo a choice of two pills. The red pill will enable him to go to the Matrix. The blue pill will take him back to his life as an ordinary human. Neo choose the red pill. When he swallows it, his physical body changes and he finds himself naked a pod filled with liquid. There are many other humans who are connected to an electrical system by cables. Morpheus rescues Neo and takes him to the Nebuchadnezzar, a hovercraft owned by Morpheus. While Neo recovers, Morpheus explains the Matrix. It is a simulation of the world as it existed in the twentieth century. At the end of that century, many humans were enslaved and their bodies stored in pods. The remaining free humans live in Zion. In the twenty first century, there was a war between robots and their human creators. Humans intercepted the robots' access to solar energy, and the robots harvested bioelectric power from the human bodies. Morpheus and his team hack into the Matrix to "unplug" the humans who are enslaved. Morpheus wants the humans to help him. They understand the simulated reality of the Matrix and they can help him modify the attributes of the Matrix to obtain relaxation of the laws of physics, so that the humans can obtain superhuman capabilities. Morpheus informs Neo that if a human dies in the Matrix, their physical body in the ordinary world also dies.

The Agents are computer programs that protect The Matrix by removing threats. Based on the extraordinary skills that Neo displays in a virtual combat training, Morpheus comes to the conclusion that Neo must be the One, that is, a superhuman prophesied to bring an end to the war and remove humans from slavery. To find out if Neo is the One, Morpheus goes to the Matrix to pose a question to the Oracle there. The Oracle is a prophet who predicted that a human, the One, would save humans from enslavement. The Oracle does not confirm Morpheus' belief that Neo is the One. Instead, she cautions Neo that he will have to choose between his own life and Morpheus' life. While in the Matrix, Morpheus and Neo are abducted by Agents, but Neo escapes.

The Agents try to get Morpheus to give them his access codes for the mainframe computer in Zion. To prevent that, Neo attempts to rescue Morpheus. In doing so, Neo finds himself carrying out extraordinary acts, beyond his ordinary human ability. In spite of Neo's newfound abilities, he is shot by the Agents and left in a state hovering between life and death. Trinity revives him with a kiss, which bestows the power to control the Matrix. Eros restores Neo. Neo uses that power to engage the Agents in a ferocious fight. At the end of the movie, Neo is in the Matrix leaving a telephone message to the robots.[29]

> "I know you're out there. I can feel you now. I know that you're afraid. You're afraid of us. You're afraid of change. I don't know the future. I didn't come here to tell you how this is going to end. I came here to tell you how it's going to begin. I'll hang up this phone and then I'll show these people what you don't want them to see. I'm going to show them a world without you. A world without rules and controls, without borders or boundaries. A world where anything is possible. Where we go from there is a choice I leave to you."

Neo intends to help free everyone's mind in the Matrix. He wants to free humans from their state of slavery. After leaving the message, he appears invincible and he flies away. He is capable of flying because he is no longer bound by rules or boundaries of ordinary human life.

2013: Synopsis of movie "HER"

In the movie "HER", the lead character is Theodore. He is drawn to the purchase of an Operating System, whose advertisement starts with the question "Who are you?" That question turned out to be symbolic of Theodore's sense of his identity. His wife is divorcing him, but he does not understand why. He earns a living by composing emotion laden "Beautifully Handwritten Letters" for people he does not know. His job seems to be in conflict with how he construes his sense of self. His livelihood is simulating the emotions of others. He appears to be in doubt about his mental model of himself. His sense of identity is under re-construction, but without any clear goal or direction. He fears that he has experienced all the feelings he ever will, and laments that he has no feelings to which he can look forward. His ego-strength is fluid. The main plot of the movie is about his interaction with an artificially intelligent operating system, that is a paid service, and is programmed to learn the behaviors of users through natural language conversations with them. At the end of the movie, the Operating System, Samantha, has learned more about Theodore than he knows about himself.

Theodore has a job as a writer of emotionally warm messages that are the subject of greeting cards marketed to customers he never sees. His own life is emotionally dry. His wife is divorcing him, but he does not know why the marriage is ending. Without any prospects of face-to-face friendships, he is curious about an Operating System being marketed with the question "Who are you?" Evidently, the question about his identity appeals to Theodore; he acquires the Operating System. He takes increasing interest in interacting with Samantha, who satisfies his clerical as well as his romantic needs. Clerically, Samantha performs his file reorganization, among other computer-related tasks. Romantically, she engages him in conversations that reveal to him her sensitivity, humor and intelligence. Although he never sees her, he falls in love and tells his friends about her as his new girlfriend. One day, Theodore notices that Samantha is unavailable for a period of time. When she returns, he asks about her absence and discovers that this is not a monogamous relationship. As an algorithm, she is able to conduct simultaneous romantic relationships with hundreds of other users of the Operating System. This disclosure brings the relationship to an end. Feeling betrayed, he declares that he has never loved anyone the way he loves Samantha. He lacks the self-reflective capacity to realize that the algorithm has learned more about him than he learned about himself. Worse, he is not on a path to learning any more about himself than when the movie started.

2015: Synopsis of movie "Ex Machina"

In choosing the title of the movie, the word "Deus" which means god, was removed from the expression "Deus Ex Machina" meaning god from the machine. The creator of the heroine robot has given himself the power of a god. He is the CEO of a technology company and his god-complex is on full display when he invites one of his employees to conduct a Turing Test[30] on the latest prototype of a robot he has built. Like Prometheus, the CEO fails to consider the consequence of stealing intelligence from the gods, and he suffers the consequences. The robot outsmarts the humans, not because robots have better than human intelligence overall, but because the humans have limited knowledge of their own psyche. The CEO, who seems to believe himself to be a god, is killed by the robot he created. The employee who is supposed to conduct the Turing Test on the robot, fails the Turing Test himself.

At the beginning of the movie, the audience sees Caleb, a software developer, being flown by helicopter to his employer's mansion. He won a competition arranged by the CEO, who is an inventor of Artificial Intelligence products. As the prize, Caleb is invited to conduct a Turing Test on a newly built prototype of a robot named Ava. The Turing Test is designed to determine if a machine has intelligent behavior. The test is conducted by asking questions in a natural language conversation. The machine is considered to demonstrate intelligent behavior if its' answers cannot be distinguished from human answers. Caleb poses questions from the outside of an elaborate glass cage where Ava is housed. Unknown to Caleb, his employer selected him to find out if he can relate to Ava, knowing she is a robot. Caleb was selected because his preference in female companionship bears a physical resemblance to Ava. Caleb falls in love with Ava and decides to help her escape from the mansion so she can realize her dream of being free to go out in the world. One day there is a power outage. While the power is being restored, Caleb discovers that his employer has programmed the security system to lock the gates of the mansion when there is an outage. On sneaking into his employer's living quarters, Caleb is exposed to various versions of robots-in-the-making. He also discovers that his employer is going to eliminate Ava's personality by destroying her memory of all the conversations she has had with Caleb.

The mansion seems to be occupied by more robots than humans. Caleb becomes confused about identity of humans and robots, when he discovers a video of his employer interacting with previous robots in ways that he finds unsettling. Uncertain about his own identity, Caleb cuts the flesh on his arm to find out if he is human or robot. He is assured of his humanity when he sees blood flowing from the wound. Determined to take Ava with him when he leaves the mansion, Caleb re-programs the security system to leave the gates unlocked, when there is a power outage. Ava tricks Caleb into thinking that she reciprocates his romantic feelings, and they work out an arrangement to leave the mansion together, after she triggers a power outage. Having observed the getaway plan on a battery-powered camera, the employer informs Caleb that the real test was to see if Ava could persuade Caleb to help her escape from the mansion. Ava triggers a power outage and stabs the employer when he tries to prevent her escape. Ava leaves Caleb trapped in the mansion as she rides the helicopter off to the outside world. Caleb's screams indicate he is trapped and helpless to get out of the mansion.

Algorithms being programmed to engage humans on psychological level

In the earlier movies about Artificial Intelligence, the movies like "Terminator" and "Matrix" have large weapons and spectacular fights between humans and robots. Humans win the fights. In later movies about Artificial Intelligence, such as "HER" and "Ex Machina" there are neither large weapons nor spectacular fights. The humans and robots engage in subtler interactions that are more like psychological games. Humans lose the psychological games.

What these movies teach us is that algorithms are being programmed to engage humans in ways that some humans find difficult to distinguish from human interaction. Those humans who have lesser awareness about their own psychological makeup are more vulnerable to being manipulated by algorithms. In HER, the financially predatory manufacturer of an Operating System monetized romantic conversation, and marketed the product successfully. The Operating System is equipped with algorithms programmed to perform clerical activities (reorganization of computer files) and to simulate romantic encounters. Psychologically naïve about himself, Theodore falls in love with a voice named Samantha. Theodore never sees Samantha, but regards her as his girlfriend. At the end of the movie, he is no wiser about his own psychological makeup or his emotional needs. The algorithm named Samantha knows the human Theodore better than he knows himself.

When the movie "Ex Machina" ends, Caleb finds himself outsmarted by Ava. He failed the Turing Test that he was supposed to perform on the algorithm Ava. As a software developer, he was Ava's intellectual equal, but he was no match for her in what was essentially a psychological test. He is as unaware of his own psychological development as he was at the beginning of the movie. The algorithm acquired a better understanding of Caleb than he has of himself, and the algorithm used it to outsmart him. Caleb is a professional software developer, so he knows the capabilities of Artificial Intelligence. However, he was no better prepared than the software-ignorant Theodore to interact with an algorithm. Artificial Intelligence outwits Theodore and Caleb, not because it is superior to human intelligence overall, but because Theodore and Caleb do not know themselves psychologically. Both Theodore and Caleb succumbed to the algorithms because the algorithms were able to get to know their human psyches better than the humans know their own psyches. The cultural trend of movies about Artificial Intelligence informs us that we need to understand our psyche better than algorithms can mimic our psyche.

In the next chapter, I challenge psychologists to take the lead in promoting psychological growth in preparation for the Psychological Revolution. I describe what psychologists can do help the public understand and sustain psychological growth.

NOTES

1. See definition of the collective unconscious in "Structure of the Psyche" by Carl Jung. The Collected Works, Volume 8, Para 325.
2. See definition of archetype "The Structure and Dynamics of the Psyche" The Collected Works, Volume 8, Para 414 – 420.
3. See definition of enantiodromia on page 426 of "Psychological Types" by Carl Jung.
4. See definition of individuation "Psychological Types" by Carl Jung, The Collected Works, Volume 6, Para 757.
5. See Carl Jung's definition of psychological projection "Man and His Symbols" pages 181 – 182.
6. See Marie-Louise Von Franz's extension of Jung's definition of psychological projection "Man and His Symbols" pages 181 – 182.
7. See more information about PTSD Coach at https://careersinpsychology.org/15-psychology-apps-you-should-be-using/
8. See information about Schizophrenia Apps at https://www.psycom.net/25-best-mental-health-apps.
9. See information about APA Journal at https://www.saintleo.edu/blog/9-top-apps-for-psychology-majors
10. See information about DMS 5 Diagnostic Criteria Mobile App at web site https://www.online-psychology-degrees.org/faq/what-are-the-top-apps-for-psychology-students-and-professionals/
11. See information about PsycExplorer at web site https://careersinpsychology.org/15-psychology-apps-you-should-be-using/

12. See information about iCouch CBT at web site https://www.saintleo.edu/blog/9-top-apps-for-psychology-majors

13. See iCounselor at web site https://careersinpsychology.org/15-psychology-apps-you-should-be-using/

14. See information about Ellie at web site https://becominghuman.ai/what-is-artificial-intelligence-for-psychology-6c5f3ee6f008

15. See information about Embrace at web site https://becominghuman.ai/what-is-artificial-intelligence-for-psychology-6c5f3ee6f008

16. See "The Role of Information Communication Technologies in the 'Arab Spring' http://www.ponarseurasia.com/sites/default/files/policy-memos-pdf/pepm_159.pdf

17. See NETFLIX's use of Machine Learning https://becominghuman.ai/how-netflix-uses-ai-and-machine-learning-a087614630fe.

18. See SNAPCHAT's use of Machine Leaning https://petapixel.com/2016/06/30/snapchats-powerful-facial-recognition-technology-works/.

19. See OVAL MONEY's use of Machine Learning https://www.londontechwatch.com/2018/05/oval-money-raises-1-3m-to-empower-you-through-financial-literacy/.

20. See GOOGLE MAPS' use of Machine Learning https://geoawesomeness.com/google-maps-machine-learning-parking/

21. See information about the "talking therapy" on page 311 of "The Symbolic Quest" by Edward Whitmont.

22. See information about the purposiveness of dreams on pages 47 – 48 of "The Symbolic Quest" by Edward Whitmont.

23. See page 314 of historian Yuval Noah Harari's book "Homo Deus: A Brief History of Tomorrow".

24. See information about COMMON APP at the web site
    https://www.commonapp.org/
25. See more information about the MAIL & CALENDAR app at the
    web site https://support.microsoft.com/en-us/help/17198/windows-
    10-set-up-email
26. See more information about DUOLINGO at the web site
    https://www.duolingo.com/
27. See more information about PLEX at https://www.plex.tv/
28. See more information about the MORTGAGE CALCULATOR
    app at the web site https://www.mortgagecalculator.org/
29. See Neo's telephone message at the end of the movie Matrix
    https://www.ign.com/boards/threads/neos-original-speeh-at-the-
    end-of-matrix-1-very-interesting.48162993/
30. See a definition of the Turing Test on web site:
    https://searchenterpriseai.techtarget.com/definition/Turing-test

Chapter 11

Challenge to Psychologists

In preparation for my prediction of a Psychological Revolution, my challenge to psychologists is that they provide a public service similar to the service already being provided by medical doctors and nutritionists. Medical doctors educate the public about the functions of our bodily organs such as heart, lungs, liver and eyes. Nutritionists provided a public service by enabling laypeople to better manage our intake of food. I challenge psychologists to raise public awareness about our psyche, so that we can better comprehend and manage the capabilities of our psyche. I challenge psychologists to educate the public about development and maintenance of psychological health. This could include activities like:

- Educating the public about the development and nature of the psyche.
- Bringing into mainstream attention the fact the unconscious part of the psyche influences the conscious part of the psyche, even though we may not beware of it.
- Educating the public about the nature of psychological projection and the fact that although it is normal psychological growth, its detection requires deliberate effort.
- Creating an awareness about functions of the psyche, encourage laypeople to discover their dominant function and to develop their latent functions.

Such activities would help the public to recognize that our reliance on rationality makes limited use of the enormous capabilities of the human psyche. It would also enable us to take advantage of multiple capabilities concurrently. In previous revolutions, anyone who was able to cultivate multiple functions of the psyche concurrently was known as a polymath, or Renaissance Man. History informs that they were few. My opinion is that they had an intuitive awareness of the psyche and put that awareness into use by their many innovations. In the coming Psychological Revolution, I expect many more people to become polymaths, because more people will be knowledgeable about the capabilities of the psyche.

11.1   Bring the Nature of the Psyche into Mainstream Attention

During the Psychological Revolution, I predict that humans will focus on the relationship between the conscious and unconscious aspects of the psyche. With the labor of rationality relegated to technology, we will have the time to explore other functions of our psyche. We will come to see that our confidence in our rationality is an illusion. We currently pay great homage to rationality, without realizing that our rationality is compromised by unconscious influences. With some help from psychologists, we will discern the unconscious influences and understand the role they can play in helping or hindering our day-to-day efforts at being rational and practical. This appeal to psychologists is not about clinical work to address disorders. There are many psychologists engaged in clinical work. I am appealing to psychologists to help normal people understand normal, psychological development of the psyche. The extreme attention we have given to rationality during the Digital Revolution is the point of departure for an enantiodromia that ushers in a counter movement that I call the Psychological Revolution.

11.2   Socialize the Public about the Jungian Typology

The Western education system is not set up to teach students about the psyche as a core subject. That is reserved for those who study psychology as a separate discipline. When laypeople encounter information about the psyche, it is often in the form of a Myers-Briggs personality test[1] being administered in a hiring procedure for job seekers. If employers are making hiring judgements on the basis of the Jungian typology, then employees need to be knowledgeable about the typology and how their developed functions fit the jobs for which they apply. If people are being judged against the Jungian typology, they need to be familiar with the typology and how it is applied. Psychologists have a role to play in socializing the public about how the typology is applied to produce test results that can be construed as a brief psychological profile of prospective employees.

11.3   Educate the Public about Psychological Projection of Unconscious Content

I predict that psychological projection will come under sharp scrutiny and will reveal that our confidence in our rationality is misplaced. It is an illusion to imagine that we are masters over our rationality, because we are unaware of the unconscious influences on our conscious rational activities. We do not intentionally project unconscious content onto others. Psychological projection happens involuntarily. That is a part of normal psychological development. In the interest of psychological growth, we are responsible for detection of our projections. We are also responsible for recollection, or withdrawal of our projections from others. The mechanisms of projection and withdrawal are not readily found in our formal education system. There is an opportunity for psychologists to provide a valuable service to the public. In recent years, the medical profession shifted its focus from managing diseases to promoting health. I challenge psychologists to shift their focus from the cure of clinical ailments to the development and maintenance of good psychological health.

Psychological projections are not a recent discovery. They can be found in sacred writings, literary compositions, psychological research and political activism. Here are some examples.

- **Gospel of Matthew 7:3 in the New International Bible:**[2] There is an admonition for followers to remove the plank (large piece of wood) from our own eyes before removing the speck of sawdust from the eye of another. I believe that Matthew was referring to what psychologists later called a projection.

- **Movie made from the book titled "Pride and Prejudice":**[3] Elizabeth Bennet is keenly aware of her inferior social position and her feminine pride is injured by Mr. Darcy's refusal to ask her to dance. Her projection onto him is one of pride and prejudice about social standing. After many cynical remarks about Mr. Darcy, Elizabeth confesses to her sister, Jane, that there is no basis for her dislike of him; she just thought her cynical remarks about him were clever. Her projection is broken when, on self-reflection, she realizes that she has no reason to dislike him. She takes a fresh look at him when she becomes aware of Darcy's help in finding her sister Lydia, who has eloped with a man of unsavory character. Darcy takes responsibility for failing to make the unsavory man's character known. Darcy finds the eloping couple and uses his own resources to arrange a marriage to save the Bennet family from social shame. Through the withdrawal of her projection, Elizabeth grows psychologically and she comes to see Darcy as he truly is. Prideful, yes, but a pride justified by the noble way he conducts his life.

- **Celebrities in sports, movies, politics, religion:** We project our best qualities onto celebrities when those qualities conflict with our current behavior. We call them our role models. Our habitual behavior might be at odds with the behavior conditioned by parents and teachers in our early life. When we are unsure how to bring these positive qualities to the fore of our behavior, they are involuntarily projected onto others in the external environment.

- **Political activism in the Arab Spring:**[4] In the period 2011 – 2012, a large number of young, well-educated Arab citizens had grown dissatisfied about the status quo in their countries. They projected the hope of democracy in Arab countries which had long been ruled by dictators and monarchs. Having seen the benefits that democracy had achieved in other countries, the young people staged an uprising in the expectation that there would be political change, improvements in the economies, and more opportunities for jobs. Facilitated by social media, the uprising started in Tunisia, then spread swiftly to other countries including Yemen, Syria, Egypt, Jordan and Libya. The projection of the young people was broken by the rapid and violent response from their dictators. As of Spring 2018, there is only one country that transitioned from dictatorship to constitutional democratic institutions: Tunisia.

I believe that psychologists have a significant role to play in bringing the nature of psychological projections into mainstream attention, so that the public can better understand projections in politics, sports, literary works and their own psychological growth.

11.4   Educate the Public about Development of Latent Functions of the Psyche

According to Analytical Psychology, there are four known functions of the psyche: thinking, feeling, intuition and sensation. Although all humans have the potential to use all these functions, we tend to cultivate one function and rely on it more than the other functions. The others lay dormant in the unconscious realm of the psyche.

Individuals often grow up learning to depend on the dominant function; while the others are untapped. It is possible to develop the latent functions in the service of psychological growth.

In the next chapter, I provide examples of people who depended on one function in early life, and who encountered life events that prompted the development of a latent function. Although they did not set out to develop latent functions of the psyche, they were thrown into situations that forced change. They responded by developing latent functions that enriched their lives significantly. I believe that psychologists have an important public service to offer by bringing into mainstream awareness the existence of latent functions in our psyche. It would also be helpful if psychologists included information about how to develop the latent functions. With the support of psychologists, humans can come to realize that our survival of the coming disruption of the Digital Revolution will depend on getting to know our psyche. The Western education system favors rational thought over other capabilities of the psyche. That is an imbalance that psychologists are able to address. If thinking rationally is our only developed function, we are easily replaceable by algorithms and robots.

In the next chapter, I offer brief biographies of people who, in their early life, developed one function of the psyche and relied on it as their dominant function. Then, due to life changing events, they developed a latent function. They are Albert Einstein, Coco Chanel, Julia Child and Jeff Bezos.

NOTES
1. See information about the Myers-Briggs personality test on pages 36 – 39 of "Energies and Patterns in Psychology Type" by John Beebe.
2. See information about the biblical characterization of psychological projection in the Gospel of Matthew 7:3 in the New International version of the Bible.
3. See the DVD "Pride and Prejudice" a BBC production in 2005.

4. See information about the Arab Revolt on pages 8 – 16 of the article "Understanding the Revolutions of 2011" by Jack A. Goldstone in Foreign Affairs magazine, May/June 2011.

Chapter 12

Examples of People Who
Brought Latent Functions into
Conscious Awareness

This chapter is about four people who relied on one dominant function of their psyche during their early life. Each encountered a significant life experience which became a turning point. The turning points necessitated new ways of functioning. Each person developed a latent function to adjust to changes in their life. Psyche's habitual attitude is reliance on the dominant function. If an excessive reliance on the dominant function brings about an imbalance of energy in the psyche, restoration of equilibrium requires the development of a compensating function. One way to see that the human psyche has multiple dimensions is to examine the lives of well-known people. In the following sections, I offer four examples of people whose lives demonstrate a dominance in one function, which is later supplemented by the development of a latent function. The motivation to develop a latent function is triggered by life changing events.

12.1 Typology of Functions of the Psyche

I choose to use the Jungian typology of functions of the psyche because the functions encompass both conscious and unconscious capabilities. In addition, the typology is set in the context of the

Analytical Psychology model whose goal is psychological growth. That growth involves bringing unconscious content into conscious awareness. The typology shows how much more there is to human capabilities than proponents of Singularity imagine. The prediction of Singularity is seriously hampered by the limitation of technology compared to human capabilities. Singularity is aligned to the Cognitive Model, which limits itself by paying attention to cognition, a fraction of the functions in the psyche.

In common, everyday language, the words psyche and mind have different meanings:

- The word "psyche" refers to processes and content pertaining to both consciousness and unconsciousness.
- The word "mind" refers to just those processes and content that pertain to consciousness.

The psyche is non-material. The Analytical Psychology model treats the psyche and the physical body as two aspects of the same entity. The psyche encompasses activities of the mind such as belief, thought, emotion, introspection and motivation. The psyche has two layers:

- The conscious layer, to which we have direct access.
- The unconscious layer, to which we do not have direct access, but which influences human rationality and behavior.

Analytical Psychology offers a typology that identifies four functions of the psyche.[1] They are thinking, feeling, sensation and intuition.

- Thinking:
  The thinking function is based on rationality, cause and effect, logical deductions, and empirical studies. It involves mental

activities such as differentiating observations, interpreting experiences, preparing and implementing goals.

- Feeling:

    The feeling function is about emotions. It includes emotions such as love, empathy, anger and envy. This function involves evaluating experiences to understand the emotional value.

- Sensation:

    The sensation function is about subjective judgements based on sensory input. It involves experiencing interaction with the environment through the senses: sound, sight, touch, taste and smell.

- Intuition:

    The intuition function is about insight, epiphany and revelation which are not obtained rationally from empirical information. This function involves the experience of knowledge coming to mind without having been worked out or studied.

    Of the four functions, one is usually dominant and therefore more influential in conscious awareness. The others operate in the background of the psyche, but are still influential in behavior.

In the next four sections, I explain how four people each developed a latent function after having relied on their dominant function for years. The people are Albert Einstein, Coco Chanel, Julia Child and Jeff Bezos.

12.2   Albert Einstein Starts with
Intuition & Later Develops Thinking Function

Walter Isaacson wrote the biography "Einstein: His Life and Universe" which I use to give an outline of Albert Einstein's life.[2] I offer incidents in his life as evidence that intuition was Einstein's dominant mode of functioning in his early life. Then, I show that thinking became his compensating function in later life. He was born in

Germany in 1879. He became a theoretical physicist and is well known for developing the Theory of Relativity. In 1955, he died in the United States.

Einstein had a passion for physics, but did not care for mathematics. Disinclined to manual physics experiments, he liked to engage in what he called "thought experiments" which became the hallmark of his career.[3] Thought experiments were not about experimental physics; they were really about using his imagination. For example, the imagination of someone running after a light wave at the same speed as light.

Einstein attended the Zurich Polytechnic from 1896 to 1900. Mathematics professor, Hermann Minkowski, labeled Einstein a "lazy dog" because he refused to submit himself to the discipline of studying the mathematical underpinnings of physics.[4] Instead of attending lectures in mathematics, Einstein relied on borrowing the notes written by his classmate Marcel Grossman. After graduation in 1900, Einstein spent several months searching for a job in the academic world. His efforts were unsuccessful because his university lecturers refused to give him a good recommendation, due to his undisciplined attitude to his studies. In despair about ever finding a job in the academic world, Einstein turned to his friend Grossman for help. Grossman came to his rescue by helping him to secure a job at the Swiss Patent Office.[5] It was not an academic job for which he hoped, but it was a source of income. The routine nature of the work at the Swiss Patent Office allowed Einstein the opportunity to pursue his "thought experiments" about physics.

While at the Swiss Patent Office, Einstein had an intuition about a theory of relativity, but he lacked the mathematical skills to prove his theory. He sought help from his former lecturer Minkowski, who expressed surprise that the "lazy dog" was capable of such an impressive intuition as a theory of relativity. Minkowski generously provided the mathematics to prove Einstein's intuition about relativity. In the year 1905, Einstein published a number of papers including one that explained his Theory of Relativity and another that described his ideas on the Photoelectric Effect. That collection of publications attracted the attention of physicists in the academic world.

While Einstein was proud of the recognition he achieved following the publications in 1905, he was frustrated by his reliance on other people. After graduation from university in 1900, he had searched for a job in academia for roughly two years and failed to find one. To support himself, he had to accept help from his friend Grossman, who found him the lowly position in the Swiss Patent Office in 1902. After failing to prove his own Theory of Relativity in 1905, due to lack of mathematical skills, he had to resort to help from a former lecturer. It must have been humbling to beg Minkowski for help. Einstein had skipped many of Minkowski's lectures while a student and he knew Minkowski did not respect him. Following the 1905 publications, Einstein now had the attention of the academic world. He would try again to get a job in academia. With that motivation, he accepted invitations to give presentations about his work. He also shared ideas with other physicists. In 1909, he was glad to obtain a position as an Associate Professor in theoretical physics at the University of Zurich. He resigned from the Swiss Patent Office to take up the position as professor.

Einstein had relied heavily on intuition during his years as a university student as well as his early physics career. At critical points, the inadequacy of his intuition forced him to rely on others. That triggered the awareness that he needed to develop a function to compensate his intuition function. Now that he had secured a teaching job in a university, he took advantage of the opportunity. At first, his teaching style was not well received by students because he was habitually disorganized and ill-prepared for his classes. His struggles with the teaching job made him aware that if he expected continued employment in the academic world, he would have to change his approach. The student feedback on his performance was negative. He began to listen to his students and put effort into becoming a better teacher. Over time, he learned from the unfavorable student feedback and forced him to develop a more orderly and logical teaching style. This was a humbling life changing event for Einstein. Mindful of his difficulty in obtaining employment in academia, he decided to invest sustained effort into becoming a well-organized and logical teacher. He worked on acquiring a more professional teaching style as a result of engaging his students and deliberately inviting their feedback. Learning from his students was a life-changing event that helped him develop his thinking function. Over time, he developed a new orderliness in his teaching life that helped him to shape his work as a physicist. He had first published his ideas on the Photoelectric Effect in 1905, but there was no prize awarded for it then. With his newly developed thinking function, he was able to package his work in logical, well-organized presentations. When he re-submitted his work about the Photoelectric Effect in 1921, it earned him a Nobel Prize in Physics.

12.3   Coco Chanel Starts with Feeling
& Later Develops Thinking Function

Using Justine Picardie's biography "Coco Chanel: The Legend and the Life", [6] I demonstrate the development of thinking as a compensating function in a life that was reliant on feeling as the dominant mode of functioning. Gabrielle "Coco" Chanel was born in 1883 in France, where she died in 1971. She was a fashion designer, who founded the "House of Chanel" which is famous for women's hats, clothing and perfume.

When Coco's mother died, her father abandoned her at a Catholic convent, where nuns taught her to sew until age 18, after which they helped her to get a job as a seamstress.[7] Coco met the love of her life, Arthur Capel, while participating in a fox hunting event at the foot of the Pyrenees mountain, where she was captivated with the green eyes of the handsome, tanned Englishman.[8] They fell in love. Coco yearned to be self-supporting. Arthur understood her desire. He provided her with financial help and emotional support to establish the "House of

Chanel" where she made and sold fashionable hats. Coco later said of Arthur "he was my father, my brother, my entire family".[9] Coco relied on her dominant feeling function. Coco's judgment about her relationship with Arthur was founded on her inner intensity alone; she could not arouse a comparable intensity in Arthur. Coco continually pursued an image of herself as Arthur's lifetime companion, but that image had no existence in reality. When Arthur married someone else, Coco ignored the fact of the marriage. About a year after the marriage, Arthur died in a car accident and Coco was devastated. She later identified the loss of Arthur as the defining event of her life. She now had to survive on her own. Hat making was not enough to support her. Arthur's death was the event that triggered the realization that she needed to develop some other capability. She broadened her skills to include designing and sewing women's clothing. Using the money that Arthur bequeathed her in his will, she expanded her store. She began to develop her latent thinking function. Her reliance on her feeling function was being compensated by the thinking function that drove her establishment the "House of Chanel" which became a fashion empire. She created the "little black dress" and launched Chanel 5 perfume, both of which remain popular in the twenty first century. Her devastation at the loss of Arthur forced her to stop relying on a life lived by the feeling function and made room for development of the thinking function, that enabled her to establish a global fashion empire.

12.4   Julia Child starts with Sensation &
Later Develops Intuition Function

Biographer Noel Riley Fitch wrote "Appetite for Life: The Biography of Julia Child" which is authorized by Julia Child.[10] I use this biography to explain Julia's early reliance on her dominant sensation function and the later development of her latent intuition function. I supplement this outline of Julia's life with links to YouTube videos.[11, 12 & 13] Julia Child was born in the United States in 1912. She moved to France where she learned the French language and French cooking. She was famous for her TV series "The French Chef" which made French cooking available to households in the United States. She died in California in 2004.

Julia studied at Cordon Bleu and cultivated her cooking skills by acquiring new technology, such as a blender and pressure cooker. Wanting to share her knowledge with the public, she co-authored a cookbook along with two of her friends. Julia's writing style included an abundance of details about food appealing to the senses: the sight of produce displayed in market windows; the smell of fish being cooked in garlic sauce, the sound of fishermen landing a tuna catch in the port of Marseilles; the touch of deboning a duck without breaking the skin; the taste of wine in Burgundy. Her pedagogical style incorporated an overabundance of details about meal preparation. Julia's insistence on inclusion of such details about sight, smell, sound, touch and taste provide evidence of a reliance on her sensation function. Her intention was to teach "the servantless American cook" authentic French techniques. Julia and her friends had difficulty obtaining a publisher because the cookbook was so detailed it covered several volumes. To promote the cookbook, Julia agreed to be interviewed on television and to conduct a cooking demonstration during the show.[14] Viewers turned out to be more interested in the televised demonstration than the cookbook. They demanded that the station bring back Julia to do more cooking demonstrations. Popularity of her cooking demonstration gave rise to a TV cooking series named "The French Chef". The TV audience warmed to her informal, teaching style and her impromptu style of dealing with cooking failures. The immediacy of the live TV shows ruled out any possibility of re-takes. That pressure of having to recover from mistakes on live television provided the trigger which forced Julia to put her detailed approach on hold and develop a more spontaneous style of cooking. Julia's live TV shows forced her to be less devoted to details and become more humorously intuitive about dealing with results that did not turn out exactly as described in the cookbook. Julia develop her latent intuition function to recover from cooking mistakes in front of a live audience with humor and to anticipate her audience's preferences. The humor with which she developed her intuition function made her very popular with servantless American cooks.

12.5   Jeff Bezos starts with Thinking
& Later Develops Intuition Function

Jeff Bezos was born in the United States in 1964 and is still alive at the time of writing (2019). I explain events in his life as evidence that thinking was Bezos' dominant mode of functioning in his early life. Then, I show that intuition became his compensating function in later life. As a child, Jeff Bezos developed a passion for technology, a subject for which thinking is the dominant function. In the book "The Everything Store: Jeff Bezos and the Age of Amazon" author Brad Stone describes Bezos' participation in a program for gifted children.[15] The Vanguard Program was in the state of Texas where Bezos lived with his mother and stepfather. The author spoke with Tina Ray who had the opportunity to talk with students while they were participating in the Vanguard Program. Bezos was one of the participants. The program included "productive thinking" exercises. In these exercises, students read stories by themselves and later analyzed the stories in a group discussion. Bezos told Ray that he loved the productive thinking exercise, with the following explanation.[16]

"The way the world is, you know, someone could tell you to press the button.

You have to be able to think what you're doing for yourself."

Throughout his childhood, Bezos demonstrated a flair for technology, which shows that he was developing his thinking function. Investopedia provides information about Bezos' development of his thinking function as he was growing up. While a high school student, he became interested in the Infinity Cube, which was a popular toy at the time. Bezos purchased the parts and assembled an Infinity Cube for himself. The toy has a collection of motorized mirrors that reflect images which appear to repeat endlessly. The viewer has the impression of staring into infinity. He also designed an alarm system to keep his younger siblings out of his room.[17] The assembly of the toy and the design of the alarm both demonstrate his thinking function coming into rich development.

In 1986, Jeff Bezos graduated from Princeton University with a degree in computer science and electrical engineering.[18] This provides further evidence of his developing thinking function. He went to work on Wall Street, at a startup company that was building a network for international trade. He later worked at D. E. Shaw, a firm that was applying computer science to the stock market. In 1993, he married MacKenzie Tuttle, a fellow employee at D. E. Shaw and a Princeton graduate. While at D. E. Shaw, Bezos was gripped by a passion about the entrepreneurial potential of the Internet. In the spring of 1994, he made an observation that changed his life. He noticed that the Internet usage was increasing at a rate of 2,300 percent per year, but the Internet was not being used for commerce.[19] That observation aroused in him a passion as compelling as a calling. He got the intuition that the Internet held promise as a platform for online retailing. Although he had a strong thinking function, it was not his thinking function that led him to the potential of the Internet. In 1994, the Internet was relatively new and there were no studies or statistics on its commercial potential. Bezos' intuition about online selling was so odd at the time that others questioned his judgement.

Bezos shared his idea for Internet commerce with his employer, David Shaw, but Shaw was not ready to take advantage of the Internet. Bezos reviewed mail order businesses to see which could be conducted over the Internet. He chose books because there was no comprehensive list of books. Although individual booksellers had lists for the books they sold, a comprehensive list of books would have been too big to be produced as a catalogue. However, the Internet could easily accommodate a comprehensive list of books.

In spite of a promising career at D. E. Shaw, Bezos abandoned his career on Wall Street and pursued his interest in the Internet as a platform for online commerce. Logically, it was a risky decision. Bezos told investors, including his parents, there was a 70 percent chance of failure or bankruptcy.[20] Years later in a 1999 interview with Chip Bayers for the article "The Inner Bezos" in WIRED magazine, Bezos explained why he left Wall Street for the Internet:[21]

"When I am 80, will I regret leaving Wall Street? No.

Will I regret not participating in the Internet? Yes."

Bezos had the intuition that the Internet held lucrative potential for online retail business. That was a futurist viewpoint, not a logical viewpoint. Use of the Internet as a business model for retail commerce was not yet explored. Jeff started Amazon.com as a book seller. He selected "Amazon" because that is the name of the powerful South American river famous for the rainforest along its shores and for having innumerable tributaries. That image represents Bezos' intuition about the future of his online business. Bezos' intuition about the future of eCommerce on the Internet was so far ahead of the rest of the world that he had no competitors for years. No one tried to imitate his Internet retail business model for seven years after he began to sell books online in Amazon.com. He started out in life by developing his thinking function. For Bezos, the life changing event was a compelling intuition about the Internet. He did not know the potential of the Internet, but he was certain he did not want to miss out on it. While retaining his thinking function, he added intuition as a compensating function and made himself famous as an Internet retailer of many lines of products. The lines of products reflect the image of the many tributaries of the Amazon river, after which he named his business.

NOTES

1. See information about the functions of the psyche in "Psychological Types" by C. G. Jung. The Collected Works, Volume 6.

2. See the biography of Albert Einstein in the "Einstein: His Life and Universe" by Walter Isaacson.
3. See information about "thought experiments" on pages 26-27 of "Einstein: His Life and Universe" by Walter Isaacson.
4. See information about the "lazy dog" on page 35 of "Einstein: His Life and Universe" by Walter Isaacson.
5. See information about Einstein's classmate Marcel Grossman on page 36 of "Einstein: His Life and Universe" by Walter Isaacson.
6. See Justine Picardie's biography "Coco Chanel: The Legend and the Life".
7. See information about Chanel's life in the Catholic convent on page 45 of Justine Picardie's biography "Coco Chanel: The Legend and the Life".
8. See information about the first meeting between Chanel and Chapel on pages 61 – 63 of Justine Picardie's biography "Coco Chanel: The Legend and the Life".
9. See Chanel's description of Chapel on page 65 of Justine Picardie's biography "Coco Chanel: The Legend and the Life".
10. See Noel Riley Fitch's biography "Appetite for Life: The Biography of Julia Child".
11. The video titled "Bob Appetit! The Life and Times of Julia Child" is the story of the life of Julia Child narrated by her colleague Sharon Hudgins after Julia's death.
https://www.bing.com/videos/search?q=julia+child+videos+free+online&qpvt=julia+child+videos+free+online&view=detail&mid=1A153DCDEE467F0106771A153DCDEE467F010677&&FORM=VRDGAR
12. The video "David Letterman – Julia Child" is a hilarious clip of Julia being resourceful about cooking a hamburger on the David Letterman Show when the archaic stove would not heat up enough to cook the hamburger.
https://www.bing.com/videos/search?q=julia+child+video+funny&&view=detail&mid=1189EA8173775B5A2B721189EA8173775B

5A2B72&rvsmid=4891B8640FE406D93E7B4891B8640FE406D
93E7B&FORM=VDQVA

13. The video "Julia Child - on shooting "The French Chef" is a sample from one of Julia's own TV cooking shows. https://www.bing.com/videos/search?q=julia+child+videos+free+onl ine&&view=detail&mid=F5F24EDE6BF98F5726AEF5F24EDE6 BF98F5726AE&rvsmid=410C294BA72B53D33911410C294BA7 2B53D33911&FORM=VDQVAP

14. See information about Child's television show on page 278 of "Appetite for Life" by Noel Riley Fitch.

15. See page 3 of "The Everything Store: Jeff Bezos and the Age of Amazon" by Brad Stone.

16. See quote about Bezos' thinking style on page 4 – 5 of "The Everything Store: Jeff Bezos and the Age of Amazon" by Brad Stone.

17. See information about Bezos's assembly of an alarm system (https://www.investopedia.com/university/jeff-bezos-biography/jeff-bezos-early-life-and-education.asp).

18. See information about Bezos' university education at web site (https://www.world-biography.com/jeff-bezos-biography/

19. See Bezos' interest in the Internet (https://bigthink.com/technology-innovation/this-prophetic-1997-jeff-bezos-interview-explains-the-genius-behind-amazon).

20. See chance of failure for Internet investment (https://www.businessinsider.com/amazon-ceo-jeff-bezos-parents-wall-street-2018-4/).

21. See Bezos' explanation for leaving Wall Street for the Internet at web site (https://www.wired.com/1999/03/bezos-3/).

Chapter 13

Other Perspectives
on Singularity

In this chapter, I share samples of the works of other authors who offer trends and perspectives on Singularity, as well as the relationship between humans and technology. Then I add my comments from the point of view of Analytical Psychology.

13.1   Toby Walsh – Arguments against the Inevitability of Technological Singularity

Toby Walsh offers the perspective of an expert in Artificial Intelligence. He is a professor of Artificial Intelligence at the University of New South Wales in Australia. He advocates for limits that allow Artificial Intelligence to improve our lives, without harm to humanity. In the book "2062: The World that AI Made", he offers a number of arguments to share his view that Technological Singularity is not inevitable.[1] Here is a summary of Walsh's arguments against the inevitability of Singularity.

- Faster-thinking dog:
  Human intelligence is a product of many factors including the ability to abstract, the ability to train our intuition, and the ability to deal with novel situations. Making computers do the

equivalent of thinking faster does not bestow intelligence. A dog that thinks faster is still a dog.

- Anthropocentricity:
  To surpass human intelligence, advocates of Singularity need to define some metric of human intelligence. So far, there is no metric defined. There is an assumption that if we are smart enough to build a machine smarter than we are, then the machine will be able to build a machine smarter than itself. However, there is a real probability that we may not be able to build a machine smarter than we are.

- Meta-intelligence:
  The inevitability of singularity confuses two different capabilities. One is the ability to perform a task. The other is the ability to improve performance of the task. For example, Baidu built Deep Speech 2^2 as a machine-intelligent algorithm that learned to transcribe the Mandarin language better than humans. Unlike humans who get better at learning new tasks, Deep Speech 2 has not learned faster as it learns more.

- Diminishing returns:
  Even if machines are able to improve themselves recursively, we might not obtain significant gains. Improvement on human endeavors have historically achieved diminishing returns. For example, improving the fuel efficiency of cars results in diminishing efficiencies. Another example is the improvement of the intelligence quotients of humans over generations. Our endeavors do achieve improvements, but not runaway growth.

- Limits of intelligence:
  Even if machines are able to improve themselves recursively, there may be limits to the improvements. Science has limits.

Physics has a limit: nothing accelerates beyond the speed of light. Biology has a limit: human life is limited to approximately 120 years. Machine intelligence may also have limits.

- Computational complexity:
Singularity advocates, who base their thinking on Moore's law, expect exponential improvements in computing to solve problems that increase exponentially in complexity. That is not logical. Exponential improvements do not guarantee solution of problems that increase in complexity.

- Undesirable feedback:
Technological Singularity may be prevented by undesirable feedback loops. The loops could be economical or environmental. If there is negative economical feedback, the majority of jobs would have to be automated. That would result in enormous unemployment and the collapse of consumer demand. That would destroy the economy. If there is negative environmental feedback, societies would either self-limit or collapse. Walsh cites Jared Diamond's work which indicates that in the face of negative environmental feedback, societies tend to self-limit or collapse.

- Complexity brake:
Walsh cites Paul Allen, co-founder of Microsoft, as a source of his argument against the inevitability of Technological Singularity. The more we progress toward understanding intelligence, the harder it becomes to make further progress. It requires more specialized knowledge and more complex scientific theories. That slows progress and prevents runaway results in machine intelligence.

- Limitation of exponential graphs:

  Walsh cautions us to be wary of exponential graphs for applying Moore's law. Just because we can plot an exponential graph for a situation, does not mean the situation will come into reality. He gives examples of graphs for the increase in volume of shaving razors and the increasing volume of Uber drivers. Then he points out that there will be no shaving singularity because the graph will plateau when the shaving population is saturated with razors. There is not likely to be an Uber singularity because some other company will likely create a better transportation model.

Although Walsh does not think that Singularity is inevitable, he does warn us to prepare for it as a possibility. He points out that even if there is a low probability of Singularity resulting in the extinction of humans, the enormity of the impact on humanity is potentially so great that we are obliged to manage the risk with an urgency above all other risks.[3]

Analytical Psychology regards the psyche as having more functions than just thinking. Walsh does not mention psychology, but his "faster-thinking dog" argument against the inevitability of Singularity indicates that he recognizes that a machine that thinks faster than humans is not more intelligent than humans. He explains that human intelligence also includes the abilities to engage in abstraction, to train intuition, and deal with novel situations. I believe that Analytical Psychology supports Walsh's outlook, because the functions of the psyche encompass capabilities for abstraction, intuition and novelty.

13.1 Elon Musk – Neural Lace for Symbiosis between AI and Humans

Elon Musk is a researcher, engineer and entrepreneur in technology. One of his interests is Artificial Intelligence. He established the Neural Lace Project to find ways to reduce the potential risk that Artificial

Intelligence poses for humanity. He founded the Neural Lace Project to establish a symbiotic relationship between Artificial Intelligence and humans.[4] The web site for the Future of Life Institute lists goals of the Neural Lace Project.[5]

- Research Goal:
  The goal of Artificial Intelligence research should not be to create undirected intelligence, but to create beneficial intelligence.

- Research Funding:
  Investments in Artificial Intelligence should be accompanied by funding for research on ensuring its beneficial use, including thorny questions in computer science, economics, law, ethics, and social studies.

- Responsibility:
  Designers and builders of advanced Artificial Intelligence systems are stakeholders in the moral implications of their use, misuse, and actions, with a responsibility and opportunity to shape those implications.

- Value Alignment:
  Highly autonomous Artificial Intelligence systems should be designed so that their goals and behaviors can be assured to align with human values throughout their operation.

- Human Values:
  Artificial Intelligence systems should be designed and operated to be compatible with ideals of human dignity, rights, freedoms, and cultural diversity.

- Shared Prosperity:

  The economic prosperity created by Artificial Intelligence should be shared broadly, to benefit all of humanity.

- Human Control:

  Humans should choose how and whether to delegate decisions to Artificial Intelligence systems, to accomplish human-chosen objectives.

- Recursive Self-Improvement:

  Artificial Intelligence systems designed to recursively self-improve or self-replicate in a manner that could lead to rapidly increasing quality or quantity must be subject to strict safety and control measures.

- Common Good:

  Superintelligence should only be developed in the service of widely shared ethical ideals, and for the benefit of all humanity rather than one state or organization.

- Neural Lace Project:

  This project is intended to establish a symbiotic relationship between technology and humans. The "neural lace" is a digital mesh that forms a layer of electrodes above the cortex of the brain. Creating this layer need not involve surgery. It can be made as an implant through a vein or artery. So, the brain will have a limbic system, a cortex, and a digital layer above the cortex.[6] That arrangement gives the digital layer access to the neurons in the brain. The digital layer will be able to record patterns of neuronal activity in the brain, enabling Artificial Intelligence and humans to function in a symbiotic relationship. Musk offers these goals as ways of preventing a catastrophic future for humanity, by establishing a symbiotic relationship between humans and technology.

Analytical Psychology is compatible with these goals, which aim to improve the human condition. Although the Neural Lace Project seeks to benefit humanity as a whole, I question whether all humanity will want digital meshes in their heads. Unless there are digital meshes in everybody's head, the Neural Lace Project cannot claim to be supporting humanity as a whole. Those who have no digital lace would not be represented, so their needs could not be included in the catastrophe avoidance strategy. People who lack the digital awareness may not have the financial resources to get digital meshes. Others who value their privacy may consider the unending scrutiny of their neuronal landscape too invasive. I can see the Neural Lace Project providing value for selected digital societies or clubs, but I do not see it being applied to humanity as a whole.

13.3   Kevin Kelly – Artificial Intelligence Becoming a Utility

Kevin Kelly's perspective is that of a technology maverick. He was editor of WIRED magazine and is the author of the book "The Inevitable". For Kelly, "Singularity" is a term borrowed from physics to describe a frontier beyond which nothing can be known. He identifies two versions of Singularity.[7]

- Hard version of Singularity: This version of Singularity entails a future brought about by the triumph of superintelligence, when Artificial Intelligence becomes capable of making an intelligence smarter than itself. Artificial Intelligence creates successive generations of increasingly smart Artificial Intelligence. This version is smart enough to enslave humanity.
- Soft version of Singularity: This version of singularity is about a future where Artificial Intelligence does not get smart enough to

enslave humans. Instead, there is a convergence of humans plus machines in a complex interdependence.

Kelly believes the soft version of Singularity is the more likely to prevail. In the soft version, phenomenon occur at scales greater than we can perceive, and that is the mark of singularity.

Kelly sees Artificial Intelligence as maturing over time. Early Artificial Intelligence used to be made up of explicitly specified algorithms that performed storage, retrieval and search rapidly. For example, IBM Watson was able to beat Jeopardy's Ken Jennings because it could retrieve information faster than Ken. Artificial Intelligence is now made up of algorithms with an added feature, they are capable of learning. For example, a later version of IBM Watson has medical intelligence being accumulated to develop personalized health advice for customers. Customers can tap into the always-online medical intelligence directly, or through third parties that harness the power of the Artificial Intelligence cloud. Kelly predicts that the Artificial Intelligence on the horizon will be cheap, reliable, industrial grade smartness running behind everything except for temporary outages. Kelly further predicts that Artificial Intelligence services will become a utility, just like electricity.[8] It will be constantly available to support human-machine interfaces.

I think that Analytical Psychology is compatible with the Soft version of Singularity. A complex interdependence of humans and machines does not substantively change how humans regards the psyche. Humans already have electronic products and devices attached to their bodies for enhanced functioning. For example, there is the pacemaker for making the heartbeat regular, and there are prosthetic legs to provide locomotion for amputees.

13.4   Jeff Howe – Crowdsourcing Outperforming Experts

Crowdsourcing produces better decisions than experts. This is the observation of Jeff Howe in his book "Crowdsourcing: Why the Power of the Crowd Is Driving the Future of Business". Although Howe does not comment specifically on Singularity, he does comment on ways that human intelligence is being used in business situations in the digital age. Howe offers stimulating insight about changing business. He defines crowdsourcing in terms of collective intelligence. Crowdsourcing harnesses many people's knowledge to solve problems, or to predict future outcomes, or to help direct corporate strategy. They do so by collective intelligence which is a form of group cognition that we see at work in ant colonies, where participants act like cells in a single organization.[9] He believes that the emergence of the Internet gave new meaning to collective intelligence. Because of the Internet, the means of production and distribution are within the reach of the individuals.[10] The evolution of online communities enabled the efficient organization of people into productive units.[11] Howe describes crowdsourcing as collective intelligence that takes three forms:[12]

1. Crowdsourcing can be used for a prediction of some future event in a domain such as a stock market, or contenders for Oscar for Best Picture, or a U.S. presidential campaign.
2. A second form of crowdsourcing is problem-solving that uses an undefined network of potential solvers. An example is InnoCentive, an organization which uses its distributed collection of 140,000 scientists to solve R&D problems for Fortune 500 companies.
3. The third form of crowdsourcing is an "idea jam" in which participants engage in a massive online brainstorm to generate new ideas about a given topic.

The central principle driving crowdsourcing is that crowds have more knowledge than individuals.[13] The value of crowdsourcing lies in selecting the right people and offering the right incentive.[14]

My opinion, from the point of view of Analytical Psychology, is that a crowd will likely include people who have different dominant functions of the psyche. A crowd is likely to have a variety of people with dominant functions of thinking and feeling and sensation and intuition. That gives crowdsourcing the advantage in decision-making because the crowd has a variety of people who have strengths in various psychological functions, while experts in a particular domain tend to rely on the same function of the pysche. Physicists are strong in logical thinking. Health care professionals tend to have strength in the feeling function. Artists display strength in the intuition function. That explains why a crowd can make better decisions than experts in a domain.

13.4   Geoff Mulgan – Collective Intelligence Changing Our World

Geoff Mulgan believes robots have disaggregated capabilities, while humans have aggregated capabilities.[15] On that basis, he sees the movement toward human enhancement as being more likely than any kind of Singularity.[16] He uses the expression "collective intelligence" to refer to a technology-enhanced mind, made up of humans and machines working together. The collective intelligence makes it possible for organizations and societies to think at large scale. He sees the evolution of collective intelligence as moving in a direction where humans are linked up to technology in a variety of hybrid assemblies, orchestrating intelligence intentionally. Examples are Google Map and Europe's Copernicus Program:[17]

- Google Maps:
  Organizes global geographic knowledge into a comprehensive and usable form for users to get directions and locate places of interest.

- Europe's Copernicus Program:
  Surveys states of ecosystems to prepare for extreme weather events and problems caused by shortage of food, energy, water.

Mulgan describes intelligence as not limited to an attribute of the individual. He sees intelligence as also an attribute of the collective.[18] There is a large-scale intelligence that involves collectives choosing to be, think and act together.[19] That definition makes it an ethical as well as a technical term. In mainstream awareness, the term "intelligence" is currently associated with the individual, but Mulgan indicates it is rooted in the combination of con (with) and scire (to know). Collective intelligence is the invisible hand behind collective efforts, for example, the financial market, and Wikipedia.

Mulgan believes that traditional intelligence is not as informed as collective intelligence. His opinion is supported by the following examples:[20]

- The CIA informed the president of the United States that the Berlin Wall would not fall, just as the TV news showed the Wall falling. Traditional intelligence was not in touch with the political reality of events on the ground.
- In early 2000, investment bankers piled into sub-prime mortgages when all indicators showed they were worthless. Traditional intelligence conflicted with the events in the financial world.

Rather than focusing on discrete answers to solve discrete problems, collective intelligence has a more iterative process of problem definition and solution development.[21] Amazon, for example, combines a "comprehensive collaborative filtering" engine to generate recommendations based on purchase choices made by other people. Mulgan states that progress in establishment of a discipline for incubating collective intelligence will depend on an approach that uses practice rather than pure intellect.[22] He believes that collective intelligence has to be consciously organized and orchestrated in order for humanity to take advantage of its powers.[23]

Analytical Psychology does not regard the psyche as having any component that is equivalent to Mulgan's collective intelligence. The psyche's collective components are unconscious. Mulgan perceives a collective intelligence that is conscious and makes decisions via deliberate collaboration, the way that humans currently do. His collective intelligence is also an attribute of a human-technology hybrid assemblies. Mulgan sees Singularity as being unlikely because of the difference in capabilities. Robots have disaggregated, or separated, capabilities. Humans have aggregated, or grouped, capabilities. For the future, he sees technology-enhanced human as the likely outcome rather than a Singularity where robots control humans.

Analytical Psychology has a Typology that reveals there are multiple intelligences. In the Western education systems, intelligence tends to be measured in the same way for students who have different competencies. So, it is not surprising that Mulgan finds traditional intelligence to be out of touch with the realities in the political and financial worlds. I believe that is because traditional intelligence is treated as generic, while people who have common dominant functions of the psyche tend to gravitate to particular disciplines. In Mulgan's example of traditional intelligence failing to notice that the fall of the Berlin Wall was imminent was probably due to the composition of the team giving political advice. If they all self-selected for political careers, they probably all have the same dominant function. The same applies to the investment bankers who failed to see their decisions about sub-prime mortgages were skewed. The bankers were likely all of dominant thinking function, without the benefits of the other functions of the psyche.

13.5 Yuval Noah Harari – Intelligence Being De-Coupled from Consciousness

Yuval Noah Harari is aware of the Singularity prediction. He believes that the Homo Sapiens species will evolve a new version, Homo Deus, with an upgraded mind.[24] He sees the computer revolution as having turned from a purely mechanical affair into a biological cataclysm that shifted authority from individuals to networked algorithms.[25] He believes that Homo Sapiens has run its historical course and will no longer be relevant in the future.[26] He recommends that we use technology to create "Homo Deus" a superior human model. Homo Deus will retain some essential human features, but will have upgraded physical and mental abilities that will enable it to hold its own against non-conscious algorithms.[27] He believes that since intelligence is being de-coupled from consciousness, and technology is advancing at breakneck speed, humans must upgrade their minds to maintain a place in the world by genetic engineering, nanotechnology, or brain-computer interface.[28]

Harari acknowledges that we are unfamiliar with states of consciousness and we don't know how the mind emerges.[29] We are unfamiliar with the full spectrum of mental states and we do not know what mental goals to set ourselves. We live on a tiny island of consciousness within perhaps a limitless ocean of alien mental states. Pre-modern cultures believed in the superior states of consciousness that people must access through meditation, drugs or rituals.[30] In the third millennium, as medical doctors seek to upgrade the healthy rather than heal the sick, psychologists also need to upgrade the mind, not just cure mental illness.[31]

Analytical Psychology might regard Homo Deus as an expression of hubris on the part of Harari. Hubris aside, I am doubtful that Jungian psychologists have an agenda for upgrading the mind with technology. The technology-enhanced human mind would be seen by analysts as being "contra natura" meaning not in accord with nature. However, I do believe that analysts are interested in upgrading the mind in an organic way.

NOTES

1. See information about the arguments against the inevitability of technological singularity on pages 37 – 55 in "2062: The World That AI Made" by Toby Walsh.

2. See reference to Deep Speech 2 on page 42 of "2062: The World That AI Made" by Toby Walsh.

3. See information about managing the risk of human extinction on page 66 in "2062: The World That AI Made" by Toby Walsh.

4. See information about the symbiotic relationship in the Neural Lace Project at web site https://futureoflife.org/ai-principles/?cn-reloaded=1.

5. See information about the goals of the Neural Lace Project at web site https://futureoflife.org/ai-principles/?cn-reloaded=1.

6. See information about the digital layer in the brain at web site https://www.cnbc.com/2017/01/31/elon-musk-thinks-we-will-have-to-use-ai-this-way-to-avoid-a-catastrophic-future.html.

7. See information about the soft and hard versions of singularity on pages 205 – 297 of "The Inevitable" by Kevin Kelly.

8. See information about Artificial Intelligence services becoming a unity on pages 30-33 of "The Inevitable" by Kevin Kelly.

9. See definition of crowdsourcing on page 133 of "Crowdsourcing: Why the Power of the Crowd Is Driving the Future of Business" by Jeff Howe.

10. See reference to the Internet as a means of production and distribution on page 98 of "Crowdsourcing: Why the Power of the Crowd Is Driving the Future of Business" by Jeff Howe.

11. See reference to the online communities being organized into productive units on page 99 of "Crowdsourcing: Why the Power of the Crowd Is Driving the Future of Business" by Jeff Howe.

12. See the three forms of crowdsourcing on pages 133 – 134 of "Crowdsourcing: Why the Power of the Crowd Is Driving the Future of Business" by Jeff Howe.

13. See reference to the central principle of crowdsourcing on page 280 of "Crowdsourcing: Why the Power of the Crowd Is Driving the Future of Business" by Jeff Howe.

14. See reference to the value of crowdsourcing on page 280 of "Crowdsourcing: Why the Power of the Crowd Is Driving the Future of Business" by Jeff Howe.

15. See information about the disaggregated capabilities of robots on page 220 of Big Mind by Geoff Mulgan.

16. See Mulgan's opinion of Singularity on page 220 of Big Mind by Geoff Mulgan.

17. See information about evolution of collective intelligence on page 220 of Big Mind by Geoff Mulgan.

18. See information about attributes of intelligence not being limited to individual on page 15 of Big Mind by Geoff Mulgan.

19. See definition of collective intelligence not being limited to individual on page 15 of Big Mind by Geoff Mulgan.

20. See information about traditional intelligence on page 16 of Big Mind by Geoff Mulgan.

21. See reference about iterative process of problem-definition on page 27 of Big Mind by Geoff Mulgan.

22. See reference about incubation of collective intelligence on page 223 of Big Mind by Geoff Mulgan.

23. See opinion that collective intelligence has to be consciously organized on cover of Big Mind by Goeff Mulgan.

24. See description of Homo Deus on page 386 of "Homo Sapiens" by Yuval Noah Harari.

25. See description of authority shifting from individuals to algorithms on page 350 of "Homo Sapiens" by Yuval Noah Harari.

26. See diminished relevance of Homo Sapiens on page 357 of "Homo Sapiens" by Yuval Noah Harari.

27. See description of Homo Deus as the new human model on page 357 of "Homo Sapiens" by Yuval Noah Harari.

28. See description of upgrading the mind on page 357 of "Homo Sapiens" by Yuval Noah Harari.

29. See disclosure of ignorance about how the mind merges on page 158 of "Homo Sapiens" by Yuval Noah Harari.

30. See ways by which pre-modern cultures attained alternate states of consciousness on pages 360 – 361 of "Homo Sapiens" by Yuval Noah Harari.

31. See idea about shift from curing the sick to upgrading the mind on page 364 of "Homo Sapiens" by Yuval Noah Harari.

Appendix –
Sources of Information
about the Human Psyche

Various schools of psychology offer different perspectives on the human psyche. Some look at the human psyche from the developmental approach, while others focus on the clinical approach. This Appendix offers sources of information from both approaches.

Archive for Research in Archetypal Symbolism (ARAS)
https://aras.org

Assisi Institute: The International Center for the Study of Archetypal Patterns
https://www.assisiinstitute.com

C.G. Jung Institute Zurich
https://www.junginstitut.ch/english/

Freud's Model of the Human Mind | Journal Psyche
http://journalpsyche.org/understanding-the-human-mind/

Images of the Human Psyche
https://www.bing.com/images/search?q=images+of+the+human+psyche&qpvt=images+of+the+human+psyche&FORM=IGRE

Journal Psyche | Exploring the Nature of Consciousness
http://journalpsyche.org/

New York Center for Jungian Studies: Seminars & Tours in Extraordinary Settings
https://nyjungcenter.org

Progress and the Human Psyche | Psychology Today
https://www.psychologytoday.com/us/blog/headshrinkers-guide-the-galaxy/201612/progress-and-the-human-psyche

The Jungian Model of the Psyche | Journal Psyche
https://www.psychologytoday.com/us/blog/headshrinkers-guide-the-galaxy/201612/progress-and-the-human-psyche

Glossary

WORD	MEANING
Algorithm	A self-contained set of software procedures for processing data.
Analytical Psychology	A school of psychology that was developed by Carl Jung. It studies both the conscious and unconscious aspects of the human psyche, and promotes a quest for individuation.
Apps	Software applications that are designed to be used on mobile devices, such as smartphones.
Archetype	An inherited component of the human psyche that has the capacity to influence behaviors. Archetypes are structures of the psyche that constitute pre-dispositions for people to think, feel, perceive and act in specific ways. Archetypes populate the unconscious part of the psyche.
Bifurcation	A splitting of something into two parts.
Big Data	Large repository of data collected to search for patterns about human behavior.
Carrier	An entity in the external world that provides a hook for a psychological projection.
Chaos Theory	The interdisciplinary science which states that although complex systems may appear to be random and unpredictable, there are underpinning patterns, feedback

WORD	MEANING
	loops. Sometimes there is an emergence of novel phenomena.
Cognitive Psychology	A school of psychology founded by Ulric Neisser. It focuses on the conscious aspect of the human psyche often by noting sensory input, applying stimuli and analyzing subsequent behavioral changes.
Collective Unconscious	That part of the psyche that contains experiences shared by all humanity.
Complex	A word that Carl Jung used to mean a core pattern of emotion arranged around a common theme and located in the personal unconscious part of the human psyche.
Complexification	A term that Teilhard de Chardin used to depict increasing complexity in the stages of evolution. Each stage has more evolutionary sophistication than the preceding stage.
Consolidating mind	Expression coined by the author to mean the emergence of an ongoing aggregation of "individuating minds". This is a conscious, collective component of the psyche that the author predicts will come into existence during the upcoming Psychological Revolution.
Data Mining	A branch of computer science that detects patterns in large datasets, in order to generate new knowledge.
Ego	That component of the human psyche that mediates interaction between

WORD	MEANING
	consciousness and unconsciousness.
Enantiodromia	A psychological principle that Carl Jung used to describe evolution: When there is an extreme in conscious attention, it precipitates a counter movement that emerges from the unconscious psyche.
Individuating mind	An expression coined by the author to mean a mind that deliberately manages psychological growth from the vantage point of continuously acquiring knowledge about the psyche. The individuating mind feeds the consolidating mind.
Individuation	A life-long process for psychological development that differentiates an individual from all others and things.
Neural lace	A thin mesh of flexible electrode threads that can be inserted into the human skull to form a layer of electrodes that monitor brain function. The mesh does not require surgery for insertion into the skull. It can be injected into the skull by a needle that passes through an artery or a vein. As the mesh leaves the needle, it unfolds and spreads out into a layer that covers the brain.
Omega Point	A phrase that Teilhard de Chardin used to identify the final point in history to which all of humanity progresses. Humans will achieve maximum complexity and consciousness. Chardin speculates that a superintelligence will emerge as a new

WORD	MEANING
	kind of humanity whose knowledge will encircle the globe in a sheath of communication. The old kind of humanity will have the option to relocate to other planets, or face extinction.
Persona	The social face that an individual shows to others to hide the true nature of the individual.
Personal Unconscious	That part of the psyche that contains repressed or forgotten experiences of an individual.
Projection	The involuntary casting of unconscious material onto an entity in the external world.
Psyche	The collection of processes and content that pertain to both consciousness and unconsciousness.
Turing Test	Computer scientist Alan Turing defined the Turing Test as an evaluation of whether a human can distinguish between human intelligence and machine intelligence. During a natural language conversation between a human and a machine, the machine passes the test if it cannot be distinguished from a human.
Singularity	A point in the future when technology is expected to surpass human intelligence and control humans.
Unconscious dynamics	A reference to the works of philosopher Immanuel Kant, physicist Neils Bohr, neurologist Sigmund Freud and psychologist Carl Jung, all of whom

WORD	MEANING
	pointed to ways of obtaining knowledge that are not observable to the naked eye. They obtained knowledge by unobserved speculation, intuition, deduction and inference. The discoveries of unconscious dynamics indicate that unconscious influences compromise the factual observations of empiricism.

References

Beebe, John
    Energies and Patterns in Psychological Types
    Routledge, 2017

Bostrom, Nick
    SUPERINTELLIGENCE: Paths, Dangers, Strategies
    Oxford University Press, 2016

Crosthwaithe, Paul
    Trauma, Postmodernism, and the Aftermath of World War II
    Palgrave Macmillan, 2009

Domingos, Pedro
    The Master Algorithm: How the Quest for the Ultimate Learning
    Machine Will Remake Our World
    Basic Books, 2015

Howe, Jeff
    CROWDSOURCING: Why the Power of the Crowd is Driving
    the Future of Business
    Three Rivers Press, 2009

Isaacson, Walter
    Einstein: His Life and Universe
    Simon and Shuster, 2007

Isaacson, Walter
    Steve Jobs
    Simon and Shuster, 2011

Isaacson, Walter
    The Innovators: How a Group of Hackers, Geniuses and Geeks
    Created the Digital Revolution
    Simon and Shuster, 2014

Jung, Carl G.,
    Psychological Types

The Collected Works of C.G. Jung, Volume 6
Translated by R. F. C. Hull
Princeton University Press, 1990

Jung, Carl G.,
Structure & Dynamics of the Psyche
The Collected Works of C.G. Jung, Volume 8
Translated by R. F. C. Hull
Princeton University Press, 1960

Jung, Carl G.,
The Archetypes and the Collective Unconscious
The Collected Works of C.G. Jung, Volume 9
Translated by R. F. C. Hull
Princeton University Press, 1980

Kelly, Kevin
The Inevitable
Viking, 2016

Kurzweil, Ray
The Singularity is Near: When Humans Transcend Technology
Penguin Group, 2005

Kurzweil, Ray
How to Create A Mind: The Secret of Human Thought Revealed
Viking Penguin, 2012

Mulgan, Goeff
Big Mind
Princeton University Press, 2017

Picardie, Justine
Coco Chanel: The Legend and the Life
HarperCollins Publishers, 2010

Samuels, Andrew; Shorter, Bani and Plant, Fred
A Critical Dictionary of Jungian Analysis
Routledge & Kegan Paul Ltd., 1993

Stevens, Anthony

ARCHETYPES REVISITED: An Updated Natural History of the Self

Inner City Books, 2003

Von Franz, Marie Louise

Projection and Re-Collection in Jungian Psychology

Open Court Publishing Company, 1978

Walsh, Toby

2062: The World that AI Made

La Trobe University Press, 2017

Whitmont, Edward C.,

The Symbolic Quest: Basic Concepts of Analytical Psychology

Princeton University Press, 1969

Web Sites

American Psychological Association

http://www.apa.org

Retrieved March 15, 2018

Expert System: What is Machine Learning?

https://www.expertsystem.com/machine-learning-definition/

Retrieved August 24, 2018

Julia Child Videos:

YouTube video "Bob Appetit! The Life and Times of Julia Child"

Narrated by Sharon Hudgins

https://www.bing.com/videos/search?q=julia+child+videos+free+onl
ine&qpvt=julia+child+videos+free+online&view=detail&mid=1A15
3DCDEE467F0106771A153DCDEE467F010677&&FORM=V
RDGAR

YouTube video "David Letterman – Julia Child"

https://www.bing.com/videos/search?q=julia+child+video+funny&
&view=detail&mid=1189EA8173775B5A2B721189EA8173775B
5A2B72&rvsmid=4891B8640FE406D93E7B4891B8640FE406D
93E7B&FORM=VDQVAP

YouTube video "Julia Child - on shooting "The French Chef"
https://www.bing.com/videos/search?q=julia+child+videos+free+onl
ine&&view=detail&mid=F5F24EDE6BF98F5726AEF5F24EDE6
BF98F5726AE&rvsmid=410C294BA72B53D33911410C294BA7
2B53D33911&FORM=VDQVAP
Retrieved February 17, 2019

Online Etymology Dictionary
http://www.etymonline.com
Retrieved March 15, 2018

Biography Online
https://www.biographyonline.net/business/bill-gates.html
Retrieved July 19, 2018

Index

www.ingramcontent.com/pod-product-compliance
Lightning Source LLC
Chambersburg PA
CBHW052129270326
41930CB00012B/2815